The Future of Wireless Communications

For a listing of recent titles in the *Artech House Mobile Communications Series,* turn to the back of this book.

The Future of Wireless Communications

William Webb

Artech House
Boston • London
www.artechhouse.com

Library of Congress Cataloging-in-Publication Data
Webb, William, 1967–
  The future of wireless communications / William Webb.
  p. cm. — (Artech House mobile communications series)
  Includes bibliographical references and index.
  ISBN 1-58053-248-9 (alk. paper)
  1. Wireless communication systems. I. Title. II. Artech House mobile
communications library
  TK5103.2 .W425 2001
  384.5'01'12—dc21   2001022209

British Library Cataloguing in Publication Data
Webb, William, 1967–
  The future of wireless communications. — (Artech House
  mobile communications library)
  1. Wireless communication systems
  I. Title
  621.3'82

  ISBN 1-58053-248-9

Cover design by Igor Valdman

© 2001 ARTECH HOUSE, INC.
685 Canton Street
Norwood, MA 02062

International Standard Book Number: 1-58053-248-9
Library of Congress Catalog Card Number: 2001022209

10 9 8 7 6 5 4 3 2 1

To Alison—
Predicting the future of wireless may be difficult, but I can predict with confidence
that the next 20 years together will be as wonderful as the last 10.

Contents

Preface

Since 1998 I have been working in a strategic role. This has broadly required that I try to understand the future of wireless communications and develop internal strategy so that the company I work for is best positioned to take advantage of the future. There are many aspects to this, including standards, regulatory issues, spectrum, internal funding, and building an external consensus. However, the key to the whole process is the development of a view of the future.

In working on developing this view, I have noted that many who try to predict the future typically do so either in a particular area (such as cellular radio) or in numeric terms—often subscriber numbers, over time frames that typically extend out only five years from the current date. There seemed to be nobody producing a coherent and well-thought-out vision as to how the whole world of wireless communications would develop, and perhaps converge, over the coming years. Furthermore, nobody seemed to be prepared to try to predict 20 years out despite the fact that it seems likely that the third-generation mobile systems being developed today will still be in service by 2020.

This book is an attempt to fill the apparent gap in developing a picture of the future, without which it is difficult to develop strategy. It is also important to build a formal methodology for constructing a view of the future in order to ensure that as little as possible is missed. I have taken the opportunity to add quotations at the start of each chapter. These broadly relate to the possibilities and difficulties in predicting the future, as well as to some historical and future issues for wireless communications.

The forecast presented in this book is highly unlikely to be correct in every respect. It may not even be correct in most respects. Regardless, the information presented here and the thought process followed will be of value in helping others to build their own view of the future and to modify that view as circumstances change.

I am deeply indebted to those experts in the field of wireless communications who gave up their time in order to partake in the Delphi study that forms an integral part of this book. The study and its results are explained in detail. This book could never have been written without the help of Mark Birchler, Larry Marturano, Michel Mouly, Tero Ojanpera, Malcolm Oliphant, Siegmund Redl, Mike Short, Frank Yester, and all those who partook in the Radiocommunications Agency work, especially Lawrence Green, who facilitated the inclusion of the work here.

In writing this book I have drawn upon all my experience gained over many years in the industry. I have learned something from almost everyone I have come into contact with, and I thank all of those with whom I have had discussions. Special thanks are due to a number of key individuals. During my time at Multiple Access Communications, Professor Ray Steele, Professor Lajos Hanzo, Dr. Ian Wassell, and Dr. John Williams, among others, taught me much about the workings of mobile radio systems. At Smith System Engineering, Richard Shenton, Dr. Glyn Carter, and Mike Shannon provided valuable knowledge, as have contacts with a number of others in the industry including Michel Mouly (independent consultant), Dirk Munning (DeTeCon), Michael Roberts (Alcatel), Mike Watkins (Racal-BRT), Mike Goddard and Jim Norton (Radiocommunications Agency), and Phillipa Marks (Indepen). At Motorola I have had tremendous guidance from a range of individuals including Sandra Cook, Raghu Rau, John Thode, Bill Gorden, Rickie Currens, Don Willis, and others. In my work with institutions I have been privileged to work with Walter Tuttlebee, Peter Grant, Keith Jones, and many more. Many presentations and papers from those involved in the mobile radio industry have contributed to my understanding. Finally, as always, I want to acknowledge my wife, who supports all my endeavors to write books with good humor and understanding.

Note: The views and opinions presented in this book are those of the authors and not necessarily of the organizations that employ them. These views should in no way be assumed to imply any particular strategic direction or product development within the organizations thus represented.

PART.

I

Why it is worthwhile

CHAPTER

1

Contents

Why forecast the future?

"Still, you can't worry too much about the future. Life is not a rehearsal."

—*Billy Connolly (1942–)*

1.1 Why forecast the future?

I was at a conference in April 2000 when a leading industry analyst declared that five-year plans were a complete waste of time. Nobody, he argued, could possibly predict what would happen five years from now. Five months, he thought, was a more realistic time horizon over which to predict what would happen. He may have been exaggerating for effect, but in today's fast-moving Internet world he clearly had a point. Yet here is a book that attempts to look 20 years ahead in one of the fastest moving industries there is. Why do it, and what likelihood is there that it will be of any value?

In practice there are many good reasons, beyond human vanity, to try to predict what will happen over the next 20 years:

▸ Some activities, such as the international allocation and clearance of radio spectrum, can take 10–15 years to achieve. Without some view of what the future will look like, these activities can never be started.

▸ The world of communications is increasingly one of standards. The Global System for Mobile Communications (GSM) standards took eight years to complete, and networks were expected to have a lifetime of 10 years. The only way to ensure that technology will even come close to meeting needs 18 years from the time it was first conceived is to have some vision as to the requirements of the future.

▸ One of the key stimuli to developing new concepts and new ideas is the thought that they will be profitable based on a vision as to why they will meet societal needs at some point in the future. These concepts can only be envisioned in a framework of prediction.

▸ Humans generally have a fascination with what their lives might be like some time in the future and with how they can play a part in the change.

Of course, it is unlikely that a company will make a five- or 10-year plan based solely on the information presented here or in other articles predicting the future. They may, however, use the information contained herein as one of a number of inputs that contributes to their strategic thinking. Maybe this leads to a change in research or standardization programs. Maybe it leads to a change in the evaluation of the strategic importance of a particular project. Maybe it leads others to comment and criticize and as a result discuss what their vision of the future is. All of these inputs, small as they may seem at the time, gradually have an impact on the direction that the industry takes. If the only impact of forecasting the future is to make others contemplate and criticize, then that forecast has still helped shape the industry.

1.2   Why the forecast will be wrong

It has often been said that forecasting is very difficult, especially when it concerns the future. Many industry leaders have regretted forecasts that they have made. In 1944 the head of IBM forecast a world market for five computers. In the 1980s Bill Gates, Chief Executive Officer of Microsoft, proclaimed that 640K of memory "ought to be enough for anyone." Closer to the wireless industry and more recently, in 1996 a key and widely regarded industry analyst predicted that there would be 170 million lines of fixed wireless deployed by 2000. The number was closer to 2 million. These kinds of errors are amusing because individuals in such positions of influence seemingly should be able to make better assessments of the future. Many, however, make good forecasts that result in the development of successful businesses. Bill Gates was stunningly accurate in his assessment that the personal computer would be successful; his error was in the detail rather than the whole. Gordon Moore of Intel proved incredibly prophetic with his belief that microprocessor power would double every 18 months.[1] Others often note that as humans we tend to over-predict the changes that will occur in the next two years but under-predict the changes that will occur in the next five.

Of course, the most significant errors in forecasting occur when there are major paradigm shifts or changes that were not foreseen by those making predictions. Once widely understood, the market for telephone services was highly predictable until the Internet arose. The market for computers seemed easy to characterize until the transistor revolutionized their design and then when the Internet had an impact. The market for mobile telephones has actually been relatively easy to predict since around 1997, following a classical take-up curve. It is critical to meaningful prediction over the next 20 years to predict the paradigm shifts that might take place and their impact on society. If these are understood, the forecast will be close enough for current intents and purposes. If not, then this will have been little more than an academic exercise. The accuracy of these forecasts will not be 100%; it is the degree of "wrongness" that is key.

Actually, wireless communications is not the hardest of modern technologies to predict. That accolade probably belongs to the world of computing and especially those areas touched by the Internet. Because the Internet promises such a fundamental change in the way we interact, shop, do business, and live our lives, it is almost impossible to predict its impact on computers, networks, and more. Of course, wireless

---

1. Some have argued that widely publicized laws can become self-fulfilling. In this case, because the microprocessor manufacturers expect performance to increase at this speed, they tend to budget and target development efforts to achieve just such this gain on the basis that if they do not then they will be behind the market; this is why some predictions become self-fulfilling.

communications is not untouched by the Internet—far from it—but there are some aspects of wireless communications that slow the speed by which it changes. Many of these are the reasons why wireless access to the Internet today is limited. They include the following.

> The need for standards;

> The lack of radio spectrum, limiting the speed at which change can be accomplished;

> The cost of building a mobile network and the resulting time required to recoup investment;

> The relative cost of mobile communications compared to fixed communications;

> The sheer complexity of wireless communications.

This is not to say that these constraints will always exist—there may be a paradigm shift that will render one or more irrelevant and lead to a burst of change. However, it is unlikely that they will all be overturned.

As an exercise in understanding the predictability of the wireless industry, Chapter 7 takes a look at how an industry analyst might have predicted the next 20 years, 20 years ago. As will be seen, the prediction is surprisingly accurate. The key difference is in the area of subscriber numbers where the forecaster failed to predict the dramatic growth in numbers. However, in terms of network evolution, services offered, development of advanced functionality, and many more key issues, it would have been possible to make a good prediction. Chapter 7 concludes that the prediction would have been of great value to operators, manufacturers, and others involved in the industry despite the error in the subscriber numbers. Although this is no guarantee of the capability to predict the next 20 years, it does provide some reassurance that making such a prediction is not necessarily a pointless exercise.

1.3   Key factors influencing the future

The future is the combination of where we are today and the actions of millions of individuals, some far more important than others, who act in such a way as to bring about change. Predicting the future is an exercise in predicting the actions of these individuals based on guiding tenets such as human nature, free markets, and previous actions—not to mention a little guesswork. Listed below are some of the key factors that need to be examined and will form chapters in this book:

> • What might the end user want?

> • Where are we today?

> • What are the current state and trend of technology?

> • What constrains the speed with which the future can be changed?

> • Who are the likely actors of change and what are they likely to do?

> • How is strategy invented?

1.4 Methodology adopted in this book

There are many formal ways of forecasting, and whole books are devoted to the forecasting process. Most of these forecasts relate to numerical data, where, for example, the forecast would attempt to predict sub-scriber numbers. This is not something that we will embark on here—at least not in detail. The purpose of this book is to determine the overall strategic direction of the industry rather than an attempt to accurately understand the number of users. Of course, some understanding of user levels is important in order to understand whether business cases can be built for certain ventures, but a broad understanding of growth will be sufficient here.

Instead, in this book, we will be much more interested in qualitative forecasts: predictions of what types of services users will employ, how networks will be connected, what types of systems will be used to communicate in homes, and where the wireless Internet will take technological development and social usage. Formal forecasting methodologies that look at these issues are less plentiful [1]. Of note are the following:

> • *The Delphi technique.* This is named after the ancient Greek oracle at the city of Delphi. It is a method for gaining opinions from a group of "experts" who provide their opinions on a certain topic with a view of gaining an insight into future possibilities. After each person (or group, if the experts are segregated into groups) has

provided a first-pass analysis, the views are circulated among all
the experts. At the next meeting the experts discuss ideas from the
other groups. The results of these discussions are then circulated to
representatives from each group, who meet together to form a final
input. The critical factors in getting good results are selecting people who are truly experts, encouraging them to be very open with
their views, and limiting contact among the groups at an early stage
to avoid reaching a hasty conclusion.

▸ *Multicriteria analysis.* This is a technique used to try to avoid natural
bias that may occur in human decisions due to personality issues or
built-in tendencies. It is typically used in making a specific decision.
All the people who will be affected by the decision are listed and
grouped according to the manner in which they will be affected. A
person from each group is then asked to list the critical issues in
the decision. These issues are then shared with all participants
who jointly rank the importance of each issue. These issues and
weightings can then be used to quantitatively weight the different
proposals.

Here it is clear that there is no single "decision" to be made. Instead,
we are seeking future possibilities, and so the Delphi technique is most
appropriate. This technique has been used within the context of this
book. While writing this book, written views of the future were solicited
from seven leading experts[2] in the wireless communications industry,
chosen from academia and industry and from as wide a base of interests,
geography, and sentiment as possible. These views are reproduced in
Chapters 8–14. The views were then circulated among the individuals
and a summary of all the views was prepared. The experts commented
on the summary until it reached a point with which all could agree. The
final summary is presented in Chapter 15.

Knowing that our forecast of the next 20 years, provided in Chapter 17,
is unlikely to be perfect (although we hope it will be instructive),
this book places more emphasis on understanding the drivers and constraints that impact the future. For example, a free-market society rarely
produces a particular good or a service unless someone is prepared to pay

---

2. One of the "experts" was actually a group of experts in itself, assembled in London by
   the U.K. Radiocommunications Agency for a similar purpose.

for it, so an understanding of what society might be prepared to pay for is a sound basis to predict the future. Equally, fundamental laws of physics are rarely overcome, although occasionally human endeavor discovers "loopholes" in what were thought to be immovable principles.

Hence, much of this book looks at all the factors most likely to influence the future, with less space devoted to the future itself. This helps build a better understanding of many of the key drivers, most of which will remain valid whatever transpires in the future. If there are some unexpected paradigm changes, many of the observations on impacting factors will still be true but the final outcome may be different. By applying the same principles and the factors that were predicted correctly, a different view of the future can be developed.

An important point to note is that the intention in this book is not to provide a detailed explanation of all the technologies and networks that are discussed here. That would result in a massive book that would rapidly become too detailed and fail in its original concept to envisage the future. Instead, only the issues about each technology that are pertinent to a discussion of the future are raised. References are provided so that if the reader needs more detailed information in a particular area then they are able to pursue the information.

1.5 Structure of this book

This book is structured into four parts:

- This introduction forms the first part, providing an explanation of the purpose and methodology for the book.

- Part II deals with key drivers toward the future, examining what users might actually want and the current status of the industry, including current development programs.

- Part III deals with constraints and covers areas such as radio spectrum and standards that tend to slow or impact on a capability to reach the future.

- Part IV provides a forecast of the future based on the input from the previous section. It provides a forecast that might have been made 20 years ago to understand the problems inherent in forecasting. It then provides inputs from a number of experts according to the

Delphi technique before summarizing all that has come before it into a future road map and vision.

Reference

[1]   Stoner, F., and R. E. Freeman, *Management*, Fifth Edition, Englewood Cliffs, NJ: Prentice Hall, 1992.

PART

II

What drives the future

What might the end user want?

"Broadcasting is really too important to be left to the broadcasters and somehow we must find a new way of using radio and television to allow us to talk to each other."

—Tony Benn (1925–)

2.1   Why what the user wants is critical

Individuals only pay for things that they want. Of course, it does not always seem that this is the case, especially when taxes are concerned, but broadly this tenet is correct. Most commercial organizations only take a particular course of action if they believe it is likely to be profitable, which means that they expect someone to pay them for it and for that payment to exceed the cost of providing the service. Even though inventors may invent just because it gives them pleasure, their invention is unlikely to be commercialized unless people will pay for it. If we could understand exactly what it is that people will want from wireless communications 20 years from now and how much they would pay for it, then, coupled with a knowledge of technology, we could probably make a fairly accurate prediction. Equally, we could be fairly sure that if nobody wanted a particular product or service, it would not be a part of the future.

Regardless of the complex interaction between the number of people who want a service, their willingness to pay, the availability of alternative products, and the difficulty and cost of delivery of that service, we can be sure that if we could develop a complete list of what the user wanted 20 years from now, the future would be some subset of that list. Perhaps not everything on the list would be technologically possible or economically viable, but there would be nothing that would arrive in the future that was not on the list.

So far, so good. The real problem, however, is that most people cannot articulate what they want 20 years from now because they cannot imagine all the possibilities that might be open to them. Few people 20 years ago would have said that they wanted their own electronic home page that any of their friends or relatives could access in order to see pictures of them. Plenty would have said that they wanted video communications. Now, virtually nobody has video communications but many have their own home page and are willing to pay for the privilege.

The above example does not imply that understanding user requirements is a futile exercise. If an analyst had extrapolated "video communications" to "visual communications," then the willingness to pay for a Web site in the absence of video communications might have been determined. The key step is to understand, in the broadest sense, what individuals want and not be too restrictive as to the means that the desire might be satisfied. "Visual communications" is a much wider field and one that can lead to many realizations, some of which would have value, even if they were not what the individual originally anticipated. This section examines fundamental human and social requirements that might be met by wireless communications.

2.2 What are people prepared to pay for?

People, of course, pay for a lot of things. Many have nothing to do with wireless communications today nor are they likely to in the future. Others are directly related to wireless communications. It is the categories in the middle that are problematic. In general, people are prepared to pay for the following:

▸ *Basic needs*. These are the fundamental things that people need for themselves and their family to survive. At the simplest level they include food, water, shelter, clothing, and heating. At a higher level they include health care and services such as law and order. Once the very basic needs are fulfilled, people's expectations rise so that, for example, rather than any food, they want nutritious, healthy, and tasty food. Once basic shelter is provided, people want more rooms, a better environment, and storage space.

▸ *Time*. People are prepared to pay for services or products that result in their having more time to do the things they really want to do. This might include, for example, paying someone else to attend to their basic needs (such as mending their house). It might extend to paying for something that reduces the time that a task takes, such as a larger lawnmower that enables the lawn to be cut faster.

▸ *Communications*. Communicating with others is a basic part of our social fabric, and people are prepared to pay substantial sums for the ability to rapidly communicate with others, even though talking to a relative may not fulfill any basic need or result in saving time.

▸ *Entertainment/recreation*. Once the basic needs are met, people look for pleasurable ways to pass their free time. This could be playing sports, interacting socially, learning a hobby, or whatever. The activity does not contribute to basic needs and certainly does not save time.

Many other things that we spend money on come from these basic categories. For example, transportation is often a major expense item. It is related to many of the above—it enables basic needs such as allowing food to be brought back from the supermarket; it saves time by allowing essential transport to be accomplished faster; and it aids recreating by enabling individuals to move to where they want to be.

Wireless communications could be relevant to all these categories. Although at the moment, when people think about wireless communications they tend to think of mobile phones, which mostly fall into the communications category; a little lateral thinking shows that wireless can do much more. Wireless networks in the home can aid the manner in which the basic need of shelter is provided. Wireless solutions can bring substantial savings in time by enabling more rapid provision of information. Wireless can enable all sorts of entertainment such as interactive games and video distribution. When looking for areas where wireless might have an impact, we need to look at all areas of human activity. We will return to this point in Section 2.5 when we look at what people might be prepared to pay for in the future concerning wireless communications.

2.3 What they have today

Today, when people think of wireless communications, they tend to think predominantly of personal communications, that is, the process by which they exchange information with other individuals. Communications today is a mixed and rather disorganized environment. The typical office worker in a developed country currently has a wide range of different communications types:

- ▶ The office telephone, used mostly for voice communications, complete with mailbox system;

- ▶ The office fax machine, which is now being used less frequently as e-mail takes over;

- ▶ The office local area network (LAN), which provides high data rate communications such as e-mail and file transfer;

- ▶ Dial-up networking when out of the office, which provides the same capabilities as the LAN but at a much slower rate;

- ▶ Mobile telephones, which provide voice communications, a mailbox, and in some cases low-speed data access;

- ▶ A pager, which provides one- or two-way messaging;

- ▶ A home telephone, which provides voice communications and dial-up access along with a home answering machine;

- ▶ A home computer linked to a different e-mail system, perhaps using high speed connections such as asymmetric digital subscriber line (ADSL) or cable modems.

Managing all these different communication devices is complex and time-consuming. The worker who has all of these (and many do) will have five "phone" numbers, three voice mailbox systems, and two e-mail addresses. There is no interconnection between any of these devices; all the different mailboxes have to be checked separately, using different protocols and passwords. Contacting such an individual is problematic because of the choice of numbers to call, and many default to calling the mobile number as the one most likely to be answered. Although many are working on systems such as unified messaging, designed to allow all types of communications (e.g., voice, fax, e-mail) to be sent to one number, we are still some way from the ideal situation where individuals only have one "address" and all communications are unified. Effectively, there is little convergence, at least as far as the user is concerned, between all these different methods of communications.

People also utilize wireless communications, in the broadest sense of the definition, for many ad hoc tasks within their environment. For example:

- Radio garage door openers facilitate communications with the garage door without needing to leave the car.

- Keyless entry systems unlock cars and can perform other functions such as setting off panic alarms.

- Baby monitor devices allow the child to communicate with the parent (even if children are unaware that they are doing so).

- Television remote controllers enable users to communicate with the television without having to touch it.[1]

- Remote-controlled model vehicles and planes are used for recreation purposes.

What this category of wireless communications devices has in common with, and what differentiates it from, the previous categories of communications devices is that these devices facilitate communications between humans and machines. This is a very important and growing

---

1. This might seem a rather strange item to place in the list. Remote controllers, however, very much fall into the category of wireless communication devices, albeit using the higher parts of the radio spectrum.

area that people are prepared to pay for because it saves them time or brings them leisure. It is also worth noting that in most cases these devices are unifunctional. That is to say that the garage door opener performs only one function, the opening of the garage door. No other device can open the garage door. There are some early signs of consolidation; for example, some personal digital assistants (PDAs) can now double as television remote controllers, but at the moment, the cost of consolidating devices exceeds the utility that people perceive from having a single device.

Finally, there is the area of machine-to-machine wireless communications. This is a very poorly developed area at the moment. Many have postulated how the refrigerator might talk to the PDA to update the shopping list (and we will return to this topic in subsequent sections), but to date there are very few commercial instances of such communications. In Section 16.3 of this book we will look at the barriers that exist to prevent this form of communication and how and when they might be overcome.

It is important to look at all these different areas. A narrower definition of wireless communications as simply a means for person-to-person communications, as with today's cellular telephone, would miss areas where key changes in functionality and behavior might occur in the future.

2.4 What they want now

One day when I was at an airport with a colleague on the way to a customer meeting, my colleague decided that we needed to present an organizational chart to the customer but he did not have a recent one on his laptop. There then followed a long and convoluted process where he called his secretary and asked her to call some other people to obtain one. Then, because there was no other way, he found a public fax machine in the airport and after some difficulty got the chart faxed to him. On the first attempt the quality was terrible but after a number of retransmissions he finally got a paper copy. He was unable to turn this into the transparency format he really wanted, but at least he had gone some way to solving the problem. In an age of high technology it would seem that we should be able to do better than this. My colleague was quite clear on what he wanted—a LAN within the airport through which he could access his corporate LAN and electronically retrieve the information. If the airport LAN had been wireless, he simply would have had to open his laptop and the connection would have been made automatically. Then, at the customer site, he wanted a computer projection unit that he could use to make his presentation. Why not have this wireless as well, so that he could simply bring up the organizational chart on the laptop and press a button to make it appear on a screen? There are many users like this who know that what they have today is not what they really want. This section tries to consider those people like my colleague and asks more generally what they want from wireless communications.

If a forward-thinking individual, say, John Smith, was asked about how he would like to communicate, given no constraints, he might provide an answer along the following lines.

- ▸ *Video communications wherever possible.* When talking from the home or office, all communications should have the option of video links and hands-free talking to make communications as natural as possible. This may not always be appropriate, especially when mobile, but the option should be available.

- ▸ *Complete unification of all messaging.* Each individual has only a single "address"—which will typically be of the form john.smith@my-ISP.com (some further detail may be required to overcome the problem of multiple John Smiths)—to which all communications will be directed.

- ▸ *Intelligent filtering and redirection.* Upon receiving a message, the network, based on preferences and past actions and knowing the current status of John Smith (e.g., whether he is mobile or at home), will determine what to do with it. Work calls might only be forwarded during the weekend if they are from certain individuals; otherwise, they would be stored and replayed upon return to the office.

- ▸ *Freedom to communicate anywhere.* It should be possible to have almost any type of communication anywhere, although the higher the bandwidth and the more "difficult" the environment, the higher the cost.

- ▸ *Simplicity.* For example, upon walking into a hotel room, communications devices should automatically network with the hotel communications system, determine whether the tariff charged by the hotel is within the bounds set by John Smith, and automatically start downloading information, presenting it to John in accordance with his preferences. This would be achieved without wires and without "login" screens, indeed, without John doing anything. When he does have to interact with a device, he would prefer to do it via voice recognition rather than having to carry a keyboard.

▸ *Context-sensitive information.* In addition to being able to get information from the Internet on request, the Internet should provide John with information that he needs depending on his location, plans, and circumstances. For example, he should automatically be provided with travel information pertinent to his journey when en route. Simply by booking his trip on-line, his communicator should request travel information automatically before starting his journey and present him with problems as he climbs into his car (or even wake him earlier if his trip is likely to take longer!).

2.5   What they might want in the future

The example of John Smith has provided some excellent concepts and ideas, but as we shall see in Chapter 3, these are not particularly visionary. Most are simple extrapolations of what he has today toward increased utility. To be able to understand the potential for real paradigm shifts, we need to utilize a more structured methodology rather than conjecture and extrapolation. We already know that approaching the problem in terms of expressing the full set of human needs is too wide a starting point to lead to useful results. Instead, we will start from the definition of wireless communications, which is the specific topic of this book.[2]

---

2. Perhaps a wider question is whether this particular topic is a sensible one. If a poor categorization was made at the start, prediction would be very problematic. For example, predicting the future for the mobile phone would be difficult because it would soon become apparent that the mobile phone might become a mobile computer; hence, in order to provide a meaningful analysis, the categorization would need to be broadened to "mobile devices." But perhaps this is too wide. A pen is strictly a mobile device but clearly not intended to be covered in the definition provided here.

Wireless communications at face value seems to be a reasonably good categorization. However, clearly there are interactions between wired and wireless communications. Fortunately, although linked, these two areas are distinct and can be kept so for forecasting purposes.

Wireless communications is *the transmission of information from one point to another without the use of wires.* We can break this down into a number of subdefinitions.

> • *Transmission.* This is the physical act of sending information in whatever form through the utilization of electromagnetic waves, be they radio waves, optical waves, or any other part of the electromagnetic spectrum. At present, most communications take place between around 10 MHz and 60 GHz and then in the optical band, but we should not constrain ourselves to assume that this will always be the case.

> • *Information.* Information takes the form of a digital or analog representation of something that needs to be sent from one place to another. In relation to humans, information is most often thought of as speech, video, text, music, or other formats. In relation to machines, it is binary data that can lead to one machine performing an action that it would not otherwise have done.

> • *One point to another.* Each point can be either a person or a machine. (In principle we could also add animals into this category, but since neither humans nor machines can typically have any meaningful communications with an animal this can be excluded for simplicity.) This leads to the following subcategories of information exchange:

> 1. Person-to-person (or multiple people);
>
> 2. Person-to-machine;
>
> 3. Machine-to-person;
>
> 4. Machine-to-machine.

> • Any one communication stream might involve one or more of these basic types. For example, we might communicate with a machine that might send an immediate response back to us.

In order to understand what wireless communications might do in the future, we need to ask what all the possible instances of wireless communications, as defined above, might actually be, without regard to whether they are technically feasible or socially useful. We can then apply all the constraints discussed in Part III to determine which are *likely.*

What is clear from the definition is that wireless communications does not create information, nor does it create people or machines. It simply provides the means for these entities to communicate. If we could produce a complete list of all machines and all modes of person-to-person communications, we would be able to produce a definitive list of all potential communications types and would be some way toward our understanding of the future. Of course, this is difficult, in particular because new machines, as yet unimagined, may be invented in the next 20 years that may change the means of communications. For example, 20 years ago, most people had not envisioned the Internet as a global machine for the storage and retrieval of information and so could not have predicted that a specific form of person-to-machine communications—wireless browsing of the Internet—would be a key driver for wireless communications. Nevertheless, some did predict the arrival of the Internet based simply on the human need to access shared information, so the task is not impossible, just difficult.

We can also state some constraints that apply to different types of communications. Communications involving a person must be sensory. That is to say that the information sent to a person must be in a form that he or she can sense. Typically, this means video or audio information, with video information including visual representation of images or text. It could, in principle, also include touch and smell information, although it seems unlikely that this will be an important use of wireless communications over the next 20 years. Communications involving a machine are highly likely to be binary digital data formatted in a myriad of different ways appropriate for the machine. When a machine communicates with a person, it must do so through a translation mechanism that renders the machine data suitable for human sense, such as a means of generating speech information.

So, in the widest sense, wireless communications in the next 20 years might involve any transmission of information between people, between machines, or between machines and people. We could then conceive of the ultimate wireless communications *device*—defined as a machine whose primary purpose is to enable wireless communications—as being one or multiple devices that enabled a person to do the following.

‣ Communicate to any other person or group of people at any time, regardless of the location of the individuals, using any mode of communications available to humans [but probably limited to video (with speech), speech alone, or text].

‣ Communicate with any machine, potentially including machines not yet invented at the time the communicator was manufactured, in any meaningful manner for the human involving video, speech, and text.

We could conceive of the ultimate communications network as one that does the following:

‣ Supports any number of the above devices;

‣ Allows any machine to communicate with any other machine in a meaningful manner.

This categorization also includes the capability for devices to communicate with each other (if a device is defined as a specific type of machine); for example, a personal communicator device could communicate with another communications device embedded in the material of a shirt being worn by a person. It says nothing about the manner in which communications will actually be organized, whether there will be multiple networks, how large the cells will be, and so on, but neither does it restrict any representation of such a network. Equally, it does not restrict networks supporting networks, such as a fixed wireless link to a home that connects a mobile wireless communications environment within the home to the Internet.

This discussion forms a useful guideline for framing the prediction problem. We can be reasonably sure that there will be no paradigm shifts that will render this definition too narrow. Of course, at the moment the definition is rather broad and does not tell us what the communications network of 2020 is actually going to look like. However, it forms an excellent starting point to frame the problem. We can now apply information on the current position of wireless communications and the constraints that will limit the future in order to understand in more detail what wireless communications networks and devices will look like in 20 years.

Chapter 3 starts this process by examining where we are today. Chapters 4–6 look at the constraints and forces that will change this starting position and render this initial view of the future more specific.

Contents

The pieces of the puzzle

"Mr. Watson, come here: I want you."

Alexander Graham Bell (1847–1922).
First intelligible words transmitted by telephone
(March 10, 1876). Bell had just spilled acid on
his clothes and was calling for his assistant,
Thomas Watson.

3.1 The value of a good understanding of the starting position

In order to predict where we might be 20 years from now, it is essential to understand in detail where we are today, and in particular, what programs are in place that will set the direction for the next five to 10 years. As Chapter 5 explains, some processes such as standardization and spectrum allocation take from five to 15 years. So, in some cases, fairly accurate predictions can be made for up to 15 years from now by understanding where we are today and the spectrum allocation programs that have been set into place. This is sufficient to understand many of the possibilities but not necessarily the outcomes. For example, 10 years ago it was possible to predict that the Iridium satellite system would be deployed around 10 years hence. It would have been harder to predict that within less than a year of deployment it would be shut down.

3.2   Wireless communications

3.2.1   Overview of wireless

There is a wide range of wireless devices used for communications purposes. Today, these can be characterized broadly as cellular, fixed wireless, LANs, personal area networks (PANs), and ad hoc devices. There is also the somewhat related category of broadcasting. There is no guarantee that this categorization will continue to hold as correct or valuable in the future, but it forms a relevant way to examine the current situation. This section examines each of these different categories along with their respective networks and protocols.

3.2.2   Cellular, paging, and other mobile communications networks [1–3]

Although this section concentrates on the cellular system, most of what it contains is equally relevant to other mobile networks, including paging networks, private mobile radio (PMR) networks, specialized mobile radio (SMR) networks, and satellite communications networks. In this book we will have little to say about PMR and SMR networks because they are specialized systems for particular user groups and because they are unlikely to play a key part in the overall evolution of communications. This is not to understate their importance to the people who use the systems, nor to say that they do not have a future; it is just that they are niche rather than mainstream applications.

Cellular communications were just entering a new era at the time that this book was written. During 2000 a number of countries distributed licenses to operate third-generation cellular networks in the frequency bands 1,900–2,100 MHz. The first deployments were expected to occur during 2001 with commercial networks becoming widespread during 2002. Much has been written about the potential benefits of third-generation systems [3]. They are expected to bring the following:

- Higher data rates than second-generation systems, offering up to 2 Mbps in some environments;

- Higher spectrum efficiency, resulting in more calls per unit area than is currently possible with second-generation solutions;

- Innovative solutions that can more readily link to the Internet and other data networks to enable new mobile applications.

During 2000, many people were staggered by the size of the auction fees that some of the operators were prepared to pay for licenses to operate third-generation systems. Second-generation systems have been evolving to the point that they too can offer substantial data rates, their spectrum efficiency is much improved from the earliest deployments, and by using solutions such as general packet radio service (GPRS) and enhanced data rates for global evolution (EDGE), they too can provide innovative data solutions.

Many other developments were taking place in the cellular industry in 2000, including:

- Increasing penetrations, to levels where there were more mobile phones in many countries than fixed phones;

- Decreasing revenue per user as more and more "low-usage" individuals purchased cellular phones and as the cost per minute of a call fell with increasing competition;

- The move toward "virtual operators" where some companies marketed a cellular service but did not own a network. Instead they bought wholesale network capacity from "network operating companies" and resold it to subscribers. In this manner, companies with expertise in building and running networks could concentrate on this area. Companies with expertise in brand awareness, marketing, and dealing with subscribers could concentrate on this area. One of the leading examples of this in 2000 was Virgin in the United Kingdom providing services by leasing network capacity from One 2 One;

- An increase in the desire to enter the "mobile Internet" era by providing access to the Internet and related services from mobile phones;

▶ The first signs of operators attempting to use location information in order to provide additional value-added information to
subscribers;

▶ Discussions as to what role the cellular phone should play inside
the building and whether there should be separate picocellular
networks within the building.

A number of conclusions can be drawn from these trends as to the
current state of the industry and its future direction.

Subscriber numbers cannot grow indefinitely.   Much of the strategy in
the cellular industry to date has been based on the rapid growth of subscriber numbers. Operators have been measured in terms of the quarterly
additions to their networks. This cannot continue indefinitely. It is not currently clear when penetration levels will tail off. In most developed countries, fixed phone penetration is around 60%, and mobile growth certainly
seems to have slowed in the countries where it has exceeded this level.
Because there are some members of society that will not have a mobile
phone in the foreseeable future, such as those aged under five or over 90,
or the homeless, it would appear that 100% penetration cannot be
reached. It is possible, indeed likely, that individuals may have more than
one phone, perhaps a different one for the week and the weekend, or one
built into the car and another one portable. Such a trend would certainly
result in the number of subscriptions being greater than 100%, although
the number of subscribers would be unable to climb above a level around
the 80% mark. One subscriber, however, can use only one phone at any
time, so really it is the number of subscribers, rather than the number of
subscriptions, that is important. In this case, the growth that has seen penetration levels increase by around 5% per year in some countries only has
about four years to go before saturation is reached. At this point, the cellular industry will start to look elsewhere for its growth. Table 3.1 provides
predictions as to the future size of the cellular marketplace from 2000 to
2005.

Table 3.1 shows that by 2005 the annual growth in subscribers will
be around 14%, substantially less than the 50–100% currently experienced by some operators. It also predicts a relatively slow uptake for
third-generation systems, so that by 2005 only around 6% of cellular
subscribers will be on next-generation systems.

Table 3.1
Cellular Subscribers Worldwide by Technology (in Millions)

Technology	2000	2001	2002	2003	2004	2005
Analog	91	84	81	75	67	55
GSM	384	497	596	690	783	891
CDMA	83	115	151	182	217	252
TDMA	61	93	129	161	192	222
PDC	49	56	60	63	64	65
Third-generation	0	1	6	23	46	83
TOTAL	669	849	1,026	1,195	1,371	1,569

Source: IMS, 2000.

The revenue trend has to be reversed.   Average revenue per user (ARPU) has been falling for some time, even though average minutes of use has been generally increasing. This is due to a number of factors:

▸ The newer subscribers tend to be the less affluent members of society who typically buy a phone on a prepaid system and then make very little use of the phone, restricting it to emergency calls. The net impact of this is to reduce the average revenue seen across all users.

▸ Competition has rapidly reduced the price per minute of making a mobile phone call to the extent that it is now in some cases cost-competitive with fixed-line prices.

▸ Interconnect revenue (i.e., the revenue received by the cellular operator from another operator when a call was made to a cellular subscriber by a subscriber connected to a different network) is falling. Interconnect revenue was once a major source of profitability to cellular operators, accounting for 25–50% of their revenue. However, many regulatory bodies, including the European Commission, have decreed that such charges have been set at unreasonable levels and have mandated that they fall over time. This is having a significant effect on the operators' profit.

Operators are well aware of the increasing problems this is causing to their profitability at a time when they are looking to invest substantial sums of money in next-generation licenses and networks. They are looking for ways to overcome this problem including:

- Promoting the use of the mobile phone as the home phone: This would increase minutes of usage, and in principle cause money that would have been paid to the fixed line operator to be paid to the mobile operator. This would be a slow process since most people have always had a phone in their home and wouldn't choose to give it up for a mobile phone that may seem less reliable and of a lower quality. Nevertheless, progress can be expected in this area in coming years.

- Increasing data usage: Mobile data is seen by many as the next great revenue stream. Mobile data typically means accessing the Internet from a mobile phone or using the mobile phone for electronic commerce (e-commerce, or when used from a mobile phone, sometimes termed m-commerce). The difficulty here is to understand where the revenue really comes from. There are three key sources: (1) the airtime usage for which the subscriber can be charged, (2) revenue from advertisers who pay to have their message passed to the mobile subscriber, and (3) a transaction fee for all e-commerce transactions performed over the network. The airtime usage can clearly accrue to the operator, although this may not be particularly significant, as the data volumes transmitted are likely to be small in comparison with a voice call.[1] The advertising and transaction revenue may not accrue to the operator at all but to the owners of the sites, or portals, that the mobile subscriber chooses to visit. Although operators can build their own portals and can make these the initial default for subscribers, they cannot force subscribers to use these, and with intense competition, many will go to versions of sites like Yahoo! Mobile (mobile.yahoo.com).

- Enter the in-building marketplace so that mobile phones are used within the office instead of the office phone. This is a complex area

---

1. For Internet applications on a mobile device, the only information transmitted to the device will be that which can be displayed on the screen, typically limited text information. This requires very little data capacity, perhaps some 100 bytes—1 kbyte per "click" compared to the 1 kbyte/second (8 Kbps) typically required for voice.

where ownership of the system and connections to external networks have many difficulties. Nevertheless, some operators are deploying systems of this type.

▸ Link up with providers of other telecommunication services to provide a "bundled" offering where the subscriber can get a number of services from one source, including fixed and mobile telephony, high-speed data connections to the home, and Internet service provision. This captures increased value and perhaps loyalty for the operators and allows them to offer new services such as a one-number arrangement to the end user for which they may be able to gain increased revenue.

So, in summary, cellular operators are currently on the verge of a new world: one where there are new technologies but also new challenges in terms of increasing the average revenue per user and mastering the predicted new world of mobile data and one where cellular is moving from a high-growth industry where it was sufficient to provide voice services of reasonable quality and simply manage the growth to an industry where increasing revenue per subscriber, managing the high cost of licenses, and achieving differentiation will be important. The cellular operators of the future will act in a manner very different from those of today.

3.2.3 Fixed wireless

Fixed wireless is a complex and confusing area that is the subject of books such as [4]. The basic concept of fixed wireless is to provide a connection from the public switched telephone network (PSTN) or public data network (PDN) to the home or business. Fixed wireless might be used for the following:

▸ Simple voice communications to the home, providing a service equivalent to the fixed-line service that many developed countries already provide to most homes. This is often termed plain old telephone service (POTS) and includes voice, fax, data at up to 56 Kbps, and the provision of customized local access supplementary services (CLASS)—capabilities such as emergency calling, three-party calling, and other supplementary services.

▸ Enhanced voice and data communications, providing all of the above features, more than one simultaneous voice line, plus enhanced data capabilities. Typical offerings in this area at the moment provide data at up to 1 Mbps, but this can be expected to increase in the future.

‣ Data-only solutions for residential and business premises. These solutions are currently used for applications such as Internet and LAN interconnect. Many expect that these types of solutions will soon offer voice capabilities using protocols such as voice over Internet protocol (VoIP).

‣ High-speed and very high-speed links. These are specialized links, sometimes on a point-to-point basis, some on a point-to-multipoint basis, that provide data rates anywhere in the range 10 Mbps for basic local multipoint distribution system (LMDS) solutions to 1 Gbps for novel optical solutions.

The area is complex for a number of reasons: For each of these different types of applications there are multiple competing technologies like cable modems, ADSL modems, power-line technology, and other wireless technologies; there are multiple contending fixed wireless technologies; the business cases are little understood; and the market is highly embryonic. A summary of the segments and some of the key issues in each segment is provided in this section.

3.2.3.1 Key drivers for fixed wireless
Fixed wireless has historically been considered a means of providing fixed connections into homes and more recently as a means of local loop competition. This historic paradigm has had mixed results and has led to some failures in the fixed wireless marketplace. There is now, however, renewed enthusiasm in the world of fixed access, and a look at some of the key drivers is a good basis for understanding the current interest. The following key trends are influencing the direction of fixed wireless development:

‣ *Voice goes mobile.* Increasingly, voice calls are being made on mobile phones. Although many people today perceive mobile phones as expensive and the quality as variable, these factors will improve over time, and the increased utility gained from having a single (mobile) phone will outweigh slight cost differences between fixed and mobile. Users will decreasingly see the fixed phone as their primary voice phone but will require voice capabilities to offset lingering concerns over mobile usage.

▶ *Fixed data usage grows rapidly.* With dramatic increases in Internet penetration, data usage on fixed lines is growing quickly and by many estimates has already passed voice usage. This will continue as the Internet grows, adds video and other high-bandwidth services, and becomes more ubiquitous. Although Internet browsing from mobile devices will be possible, it will be far less satisfactory than from fixed devices; hence, the vast bulk of the data traffic will be to and from fixed devices (although there may be wireless in-home networks that allow mobility within the home).

▶ *Different users require differing amounts of data and have a differing willingness to pay.* There is already a dramatic difference in data rates required by users. Some will not pay more than the rates paid for 56-Kbps transmission. Others pay $40 per month for 1-Mbps transmission. Small businesses pay hundreds of dollars per month for 10-Mbps transmission, and large businesses pay thousands of dollars per month for transmission rates of 100 Mbps or more.

▶ *Wireless is less expensive where there is no wire.* Fixed wireless is normally a less expensive solution than installing wires (be they copper, cable, or fiber optic) to a building. Where there is wire, typically the cost of upgrading to two-way data transmission is less than or comparable to the cost of a fixed wireless solution.

These observations allow a number of conclusions to be drawn concerning fixed wireless deployments over the coming years:

▶ *Fixed wireless is a data-centered solution.* Fixed wireless growth will be more aligned with the growth in data usage, not voice, although there will be some important markets where voice is key and others where voice provides important revenue to strengthen the business case.

▶ *Fixed wireless has strong competition.* The strong demand for higher data speeds will result in a large opportunity for fixed wireless, but one for which it will have to compete with ADSL and cable modems.

‣ *Many different technical solutions will be required.* The difference in requirements and the physical characteristics of radio transmission will make it likely that there will be a number of different solutions for different user groups, each optimized to the particular characteristics of that group.

‣ *Most fixed wireless operators will be new entrants.* Key operators deploying fixed wireless will be those who do not own wired systems in the area under consideration. As a result, they are typically not the post and telecommunications operator (PTO).

From these simple deductions flow many more complicated issues. These include the extent to which fixed wireless can compete with wired solutions, the actual form of the different types of fixed wireless solutions developed, and the technological developments that might be expected. These issues are explored in the following subsections.

3.2.3.2   Key competitors to fixed wireless

There is a range of different possible delivery mechanisms for high-speed access to the premise. These include twisted pair cabling, coaxial cable (often referred to as "coax"), fiber optic cable, and even power-line cable. The key mechanisms are discussed below.

The copper infrastructure.   The copper infrastructure, comprising twisted pair copper cables and installed and owned by the state telephone company (the PTO), has formed the traditional access infrastructure for almost 100 years. Until recently it was ill-suited to anything other than voice because of its low bandwidth (only around 3 kHz). A number of recent advances, however, have started to change this. Voice-band modems can now achieve rates of up to 56 Kbps, which provides slow but viable computer access. Integrated services digital network (ISDN) provides rates of up to 128 Kbps, offering the potential of video, albeit of a low quality. The role that this copper will play in the future, though, will be dominated by the success or failure of digital subscriber line (DSL) technology, which is starting to provide a massive increase in the data rate passed through copper wire.

Research has shown that these technologies can offer anything from 1–50 Mbps depending on the quality of the existing twisted pair and the extent to which fiber has penetrated the network. The DSL technologies use advances in digital signal processing, so that although significant attenuation is experienced in the lines and although each line is different, the modems can adapt to cope with whatever environment they find.

DSL technology would, at first sight, appear to be a "killer technology" that will make all other access methods redundant. It uses infrastructure that is already in place and often already depreciated and provides data rates as high as almost any other access method. It is already providing high-speed Internet access in the United States and in some deployments has sufficient bandwidth for video on demand (VoD) services (which some operators such as BT are trialing on a small scale). There remains, however, much uncertainty surrounding DSL.

A problem with all the DSL technologies is that the data rate that can be achieved depends on the length and configuration of the twisted pair. As the length gets longer, the data rate falls. As yet, it is not clear what percentage of lines will be of sufficient quality to accept DSL signals. Figures quoted in the industry vary from around 30–90%, with this dramatic difference reflecting the lack of knowledge about the quality of the installed telephone lines around the world. Another problem is that the lines are owned by PTOs that are not known as entrepreneurs or for their speed of movement, suggesting that the introduction of DSL might be relatively slow in many countries, especially those with, as yet, limited competition. Additional difficulties are caused by a multiplicity of standards. Nevertheless, costs of DSL modems are falling quickly, and it is certain that DSL will play a major role in the future provision of broadband access.

Cable. Cable operators have implemented buried networks, which initially provided television distribution, in numerous countries. As a result, cable tends to be deployed in residential areas but typically not in business districts. Cable networks vary in their composition. Some are entirely coax. Others use fiber optic cable in the backbone but coax in the local loop; these are often known as fiber to the curb (FTTC) or hybrid fiber coax (HFC). Some have postulated fiber to the home (FTTH), but at present the economics are not favorable.

While fiber has a virtually unlimited bandwidth (on the order of Gbps), coax has a bandwidth of up to around 750 Mbps in the existing installations. With an analog TV picture requiring about 8 Mbps of bandwidth, this still allows numerous TV channels. With just one 8-MHz channel, roughly 50 Mbps of data can be transmitted. Cable, then, offers a higher capacity than basic DSL service.

This is not quite the whole story. For each premise there is one (or two) twisted copper pairs running from the switch right to the premise. In a cable network, all premises share the backbone resource and the resource of the local branch to which they are connected. To put it differently, all premises on one branch are connected to the same cable, whereas they are all connected to their own individual twisted pair. This is fine while cable is delivering broadcast services, to be watched by many simultaneously, allowing 50 or more TV channels. If each user on a branch, however, wanted a VoD service, then only 50 users could be accommodated on one branch, unlike twisted pair using DSL where as many users as homes could be accommodated. Indeed, in a typical cable network, the bandwidth per home that is available (i.e., the total bandwidth divided by the number of homes) is only around 31 kHz, although it is unlikely that all homes would be using a dedicated downlink resource at the same time. Some cable operators are working hard to "shorten" branches by running the backbone further out toward the subscribers.

This sharing of resources causes even more problems in the return direction. Not only is the return path shared among all the users who require it, significantly reducing the capacity, but also each of the users introduces noise onto the return path. The switch sees noise from across the entire network, significantly reducing the signal-to-noise ratio and hence information content that can be received. This noise is particularly severe at low frequencies where it is often known as *ingress*.

Despite all these problems, cable modems are in widespread commercial service with a downstream capability of 30 Mbps and an upstream capability of 10 Mbps, and prices are falling rapidly to below $200. Clearly, where cable networks have already been installed, cable operators are well placed to provide a convergent offering with voice, computing, and television all through the same channel. Where cable has not been laid, the costs of deploying cable could end up being higher than using fixed wireless, allowing the fixed wireless operators to undercut cable costs.

3.2.3.3 Selecting a delivery mechanism

The selection of a delivery mechanism depends largely on the existence and ownership of available paths to the home. Typically, the PTO will favor DSL, the cable operator will favor cable modems, and the new entrant will favor fixed wireless. There are hybrid situations: For example, in the United States AT&T is utilizing cable in areas where it has a cable presence and fixed wireless in other areas. Predicting relative prices and penetrations for these different delivery mechanisms is difficult because the cost, particularly for wired solutions, depends heavily on the degree of upgrade necessary to the existing network, but the current estimates from analysts are provided in Figures 3.1 and 3.2. Note that the costs for DSL and cable modems assume that the networks are already built and depreciated and already have connection to the subscribers. This may not always be reasonable—upgrade costs for both DSL and cable modems have been quoted at around $500 per household, making their overall cost very similar to high-speed fixed wireless.

Figures 3.1 and 3.2 show that there is likely to be a space in the market for each of the different delivery mechanisms. Although cable may provide the lowest cost solution, cable penetration is relatively low in many countries and further cable deployment is becoming increasingly less likely, so the momentum for cable solutions will gradually fall. DSL is likely to have the greatest number of connections but fixed wireless will fall close behind. The slight cost penalty that the high-speed fixed wireless solutions are expected to incur compared to wired solutions is likely to be overcome by a range of factors including the additional revenue that operators can achieve from the provision of end-to-end solutions, from the potential for even higher data rates that fixed wireless offers, from the ability of the fixed wireless operator to target specific and profitable niches, and from the fact that DSL can only reach a percentage of subscribers.

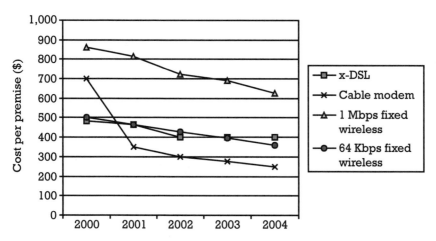

Figure 3.1   Predicted ASP per premise for different delivery mechanisms. (*Source:* Analysts' reports and Motorola data.)

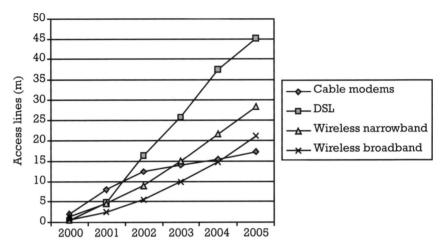

Figure 3.2   Predicted market share of the different delivery mechanisms. (*Source:* Analysts' reports.)

3.2.3.4   Segmenting the market

As mentioned in the introduction, the broadband access market consists of a range of differing requirements and willingness to pay. Equally, the characteristics of fixed wireless solutions are such that the price per premise tends to increase with the data rate provided. This is broadly because higher data rate solutions require larger spectrum allocations. These can typically only be gained at higher frequency bands. In these higher bands, the cost of radio equipment is higher and the range is lower, both resulting in a more expensive deployment. Furthermore, it becomes clear that higher bandwidth solutions require different technical choices than lower bandwidth solutions (e.g., code division multiple access (CDMA) is preferred at lower bandwidths but becomes difficult to deploy at higher bandwidths). The overall implication of this is that it is necessary to develop a range of different fixed wireless solutions for the different market segments.

By studying the requirements of the end users, the different frequency bands available, and the target markets of a number of operators around the world, it is possible to develop a market segmentation. Our current views are shown in Table 3.2.

Broadly, each of the segments detailed in Table 3.2 provides an order of magnitude increase in data rate over the previous segment for a somewhat increased cost. They span the range of frequency bands from

Table 3.2
Segmentation of the Fixed Wireless Market

	Requirements	Technology	Competition	Typical User Segment
Low-cost voice	Wireless equivalency	cdmaOne at 800 MHz and 1.9 GHz	Standard twisted pair deployment	Subscribers in developing countries that currently do not have telephones
Voice and data	2–4 voice lines and 1-Mbps data	Wideband CDMA (WCDMA) or similar at 2–10 GHz (often 3.5 GHz)	DSL and cable modems	Residential and small office/home office (SoHo) users in all countries
Data-centric	10-Mbps data	Orthogonal frequency division multiplexing (OFDM) or time division duplex/time division multiple access (TDD/TDMA) at 2–10 GHz (often 2.5 GHz)	As above	High-end residential, SoHo and small- and medium-enterprise (SME) users in all countries
High-speed	100-Mbps data and E1 emulation	Frequency division multiple access (FDMA)/TDMA at 20–40 GHz	Dedicated T1/E1 connections using copper or fiber	SMEs and multidwelling units worldwide
Very high-speed	1-Gbps data	Optical	Fiber optic cable	SMEs and large enterprises worldwide

cellular to light and the range of technologies, encompassing CDMA, TDMA, FDMA, OFDM, TDD, and more. The specific solutions discussed in each segment are described in more detail below.

3.2.3.5 Current technologies and spectrum for each of the different segments

In this section, the requirements and technologies for each of the different segments identified in Table 3.2 are discussed.

Low-cost voice.   Here the key emphasis is on the delivery of wireline transparency features (voice, analog fax, 56-Kbps data) at the lowest possible price point. In order to reduce the price, it is appropriate to adopt cellular technologies and to leverage the economies of scale that can be gained from cellular. Of all the cellular technologies, cdmaOne has the highest capacity, especially in a fixed deployment where the capacity of CDMA rises due to the "power control" problem being less severe, and is thus best suited to this particular segment.

The optimum frequency bands for this service are the relatively low-frequency bands such as 800 MHz and 1.9 GHz where equipment costs are relatively low and the propagation range is good, which minimize system cost further.

Voice and data.   In this segment the requirement is for multiple voice lines with wireline transparency as well as simultaneous data transmission with data rates of up to around 1 Mbps. A range of different technologies has been proposed for this segment. CDMA provides spectrum efficiency but with perhaps less flexibility than some other solutions. OFDM provides good rejection to multipath interference, which can be problematic at higher data rates, but at the expense of increased complexity. Either of these two solutions would be appropriate for this segment.

These solutions tend to operate in the frequency bands between 2 and 10 GHz. Below 2 GHz there is typically insufficient radio spectrum to provide users with data rates of 1 Mbps. Above 10 GHz the costs of providing radio frequency components tend to become prohibitive for this segment. The most popular frequency band worldwide for this service at the moment is the 3.4–3.6-GHz frequency band.

Data-centric.   In this segment the requirement is for efficient and higher speed data transmission, without necessarily providing full voice capability. Here, technologies that are built around packet transmission are most appropriate. CDMA technologies typically have difficulty in providing sufficiently high data rates, and instead, TDD/TDMA or OFDM technologies are better suited to the environment.

In terms of frequency band, many of the same arguments as for the voice and data segment are appropriate. However, 10-Mbps solutions require spectrum allocations on the order of 100–200 MHz, which is typically not available to a single operator in the 3.5-GHz band. As a result, data-centric solutions are currently proposed for the 2.5–2.7-GHz band—known as the microwave multipoint distribution system (MMDS) band, which is only available in North and South America—and the 5.7–5.9-GHz unlicensed band, also generally only available in the United States. The latter, being unlicensed, results in an increased risk of interference from other users but an ability to gain spectrum at no cost. It is not currently clear whether it will be possible to deploy these solutions in other parts of the world without restricting the maximum data rate that can be offered.

High-speed. This segment requires much higher speed data transmission than the other segments and as a result requires operation in higher frequency bands. Here, line-of-sight propagation and the expense of the radio frequency components tend to dominate design decisions. Again CDMA is inappropriate because of the high data rates, and instead schemes to maximize capacity on FDMA and TDMA platforms such as multilevel modulation are preferred.

To provide solutions with data rates of 100 Mbps, spectrum assignments on the order of 1 GHz are required. These are only available in the bands above 20 GHz. Assignments around the world include 24, 26, 28, 38, and 40 GHz. The fragmented nature of these assignments is somewhat problematic and can reduce the economies of scale that can be achieved.

Very high-speed. At data rates of 1 Gbps it becomes virtually impossible to provide the service without moving to laser transmission systems. Hence, solutions in this space tend to be point-to-point and point-to-multipoint laser links with relatively short ranges but very high capacity. One advantage that light transmission brings is that a license for radio spectrum is not required.

3.2.3.6 Future technologies for broadband fixed access

There is a wide range of different technologies proposed for fixed access. Here, the key technologies that we might consider in future systems are listed and briefly evaluated.

Mesh and hybrid mesh. In a mesh system, each subscriber transmits to a neighboring subscriber device, with the signal being relayed until it reaches a wired access point—a subscriber node connected into a broadband fiber optic backbone. The mesh approach effectively changes the "rules" used for capacity and links budget calculations by turning each link into a point-to-point link. Arguably, this has a similar effect to adaptive base station antennas in a conventional system, which can provide a narrow beam to each subscriber unit. The potential advantages of mesh solutions are listed as follows.

- Increased capacity as a result of frequencies being reused on a very localized level (effectively the equivalent of a microcellular approach on a conventional design, although these capacity gains could be offset by the need for each node to relay traffic);

- Improved quality as a result of each link being short and having a high link budget;

- Possible cost reduction in the subscriber unit as a result of the less demanding link budget, but this may be offset by the additional complexity required to provide the repeater element needed within the mesh architecture;

- The ability to replan the system without repointing subscriber antennas (e.g., in cases where the subscriber numbers grow more quickly than anticipated);

- The potential to have a nearly "infrastructure-less" deployment.

Against these advantages need to be balanced the potential disadvantages, which are listed as follows.

- Highly complex algorithms are required to manage the system and avoid "hot spots," which may be unstable and result in poor availability.

- Different and novel medium access control (MAC) mechanisms may be required, which will need development and will add to the complexity.

- The initial investment is relatively high since "seed nodes" have to be placed so that the mesh can form as soon as the first subscriber is brought onto the system.

- Marketing issues may be problematic in that customers may not want to rely on nodes not in their control and not on their premises for their connectivity and may not want their equipment to be relaying the messages for others.

It is difficult to draw definitive conclusions at this point since many of the above variables are unknown. If the complexity and risk can be overcome, then it seems highly likely that mesh systems will provide a greater capacity than conventional systems for a given cost and are thus likely to increase the net present value (NPV) of the operator. It is not clear whether mesh networks work best with omnidirectional or directional antennas. For the latter, they can only be deployed in higher frequency bands, which limits the applications to those requiring relatively high bit rates (perhaps 1 Mbps per user), and the need for a dense mesh limits the deployment scenarios to urban and suburban areas.

It is worth noting that the advantages of mesh networks may be frequency-dependent. At 2–3.5 GHz, conventional point-to-multipoint (PMP) systems (e.g., CDMA fixed wireless) work sufficiently well with the required infrastructure of towers and base stations. As the frequency increases, coverage begins to degrade because of the more difficult propagation conditions. It is likely that PMP will not be able to support 100% coverage to suburban areas at frequencies much above 5 GHz without numerous microcells. Above 10 GHz a mesh approach will probably be required to get suburban coverage. This increase in frequency impacts other factors, such as the amount of interference seen from other users within the mesh, additional margins for rain fading, and the cost of getting a decibel of system gain. These points would seem to recommend the use of directional antennas, which will probably be needed to improve the link budget at the higher frequencies. Both the level of interference, and the use of directional antennas will likely change the MAC layer protocol, and possibly the modulation and coding for the physical layer.

Time division duplex (TDD). Most conventional systems are frequency division duplex (FDD). Some contend that TDD brings substantial benefits to the operator. Key to this question are the following factors.

- Whether the spectrum is simplex or duplex—if it is simplex then TDD overcomes the need for a guard band, which can be wasteful of spectrum, though TDD needs a guard *time*, which may be as large percentage-wise;

- Whether the data is asymmetric and the asymmetry is time variable—if the asymmetry is known beforehand, then unbalanced FDD assignments can be used; if not, then TDD can bring some efficiency gains.

The key problem that has been postulated with dynamic TDD is that base stations might interfere with each other in an unconstrained environment. This can occur if one cochannel base transceiver station (BTS) is transmitting to a subscriber within its coverage area while a second is receiving from a different subscriber within its coverage area; because of their elevated placements there is a line of sight between the base stations. Although there is insufficient evidence to fully evaluate this problem, it would seem that the careful use of sectorization coupled with the possibility of some form of interference cancellation at base stations could be used to overcome this problem; it is unlikely to be a fatal issue for TDD.

Determining the gains is then an issue of understanding the variability of the asymmetry and the simplex or duplex nature of the spectrum. It seems likely that if the asymmetry is highly time variable, then TDD will bring definite advantages, in principle up to a maximum of a 100% capacity gain (in the case where traffic only flows in one direction—100% asymmetry). It also seems certain that in simplex bands, TDD will bring advantages unless complex frequency assignment procedures are adopted for FDD, whereby the guard band is different in different cells, requiring complex planning and possibly greater expense for the subscriber unit. Hence, assuming that the cost of implementing TDD is not great, it is likely to increase the NPV of the operators.

Forward predictions of asymmetry time variance are uncertain, so TDD gains cannot be definitively quantified, but it seems possible that there may be some worthwhile gains in certain situations.

OFDM versus single carrier.   Here, single carrier is assumed to mean existing schemes [sometimes, erroneously called quadrature amplitude modulation (QAM)], while OFDM is considered in the broadest sense to include vector-OFDM (VOFDM) and other variants. OFDM brings two benefits:

- It effectively removes the need for an equalizer by turning a wideband signal into a multitude of narrowband signals.

- It can overcome some specific types of narrowband interference more simply than other schemes.

OFDM has some disadvantages as well:

- It requires an overhead of around 12% for training sequences and cyclical redundancy (equalizers in non-OFDM solutions also require training sequences, so, depending on the size of the cyclical redundancy, this may not be an issue).

‣ Because it transforms intersymbol interference (ISI) into narrowband Rayleigh fading, it foregoes the opportunity to make use of the effective diversity in multiple paths: An equalizer will actually increase the performance by combining the multiple signals, whereas within OFDM these signals are nonresolvable and appear as Rayleigh fading.

‣ The peak/average ratio of OFDM is perhaps 3–5 dB higher than, say, quadrature phase shift keying (QPSK), putting more stress on power amplifier design. This is an issue especially for subscriber units.

It is clear that when compared to a single-carrier solution with an equalizer able to accommodate the channel ISI, OFDM will result in inferior performance. In the case where the equalizer is unable to accommodate the channel ISI, however, the single-carrier solution will typically fail, whereas the OFDM solution will continue to work. The key unknown is to what extent the channel will exhibit ISI beyond the range of a commercially viable equalizer or whether there will be narrowband interference. It is generally agreed that, to date, there is insufficient information about the channel to be able to definitively answer this problem. Given the lack of information, OFDM represents the more conservative solution, guaranteeing operation in most environments while not necessarily maximizing performance.

CDMA versus TDMA. It is generally agreed that CDMA is the most efficient multiple-access scheme for mobile applications. There are, however, different constraints within the fixed-access environment related to the desire of operators to be able to instantaneously give all, or a substantial part of the available bandwidth, to an individual subscriber. Because of the manner in which CDMA is configured, within a sector in a clustered environment typically only around 200 Kbps of throughput are available per megahertz of spectrum.[2] This compares with a QPSK TDMA solution

---

2. Recently, a number of schemes such as 1xtreme have been proposed by Motorola and others that significantly exceed this limit and move above 1 Mbps/MHz. They achieve this, however, with multilevel modulation, which increases the reuse pattern.

where up to around 1.8 Mbps might be available. Of course, CDMA allows single-frequency reuse and so the overall efficiency is high, but this example shows that in order to provide a subscriber with, say, 10 Mbps of data, CDMA would require either a carrier of 50-MHz bandwidth or a multicarrier receiver while TDMA would only require a carrier of about 6 MHz.

A further issue with CDMA is the ease of operating in a packet mode. With TDMA it is relatively simple to give a subscriber access to one or more slots if they only need to send a few packets. With CDMA there is not the same "natural" division of resources, and as a result, either a TDMA structure must be imposed or semiquiescent codesets must be provided.

It seems clear that for constant bit rate narrowband services, CDMA is the most spectrally efficient solution by some distance and, subject to the cost of competing solutions, will probably maximize the NPV for the operator. For broadband applications above 2 Mbps per user, CDMA solutions will probably not be economically viable. Below 2 Mbps, subject to there being sufficient spectrum and the cost of the CDMA system being competitive, CDMA is probably the optimal multiple-access scheme.

Adaptive antennas.   The conventional approach is to use sectored antennas at the base station, possibly with diversity, and a directional antenna at the subscriber unit, again, possibly with diversity. Adaptive antennas bring potential gains described as follows:

- At the base station they can result in a narrow beam to an individual subscriber, limiting interference to other sectors and so increasing capacity.

- At the base station they can be used to null interferers and enhance the carrier-to-interference (C/I) ratio of the received signal.

- At the subscriber unit they can be used to null interference, again increasing the C/I ratio.

It is not clear how great these gains will be. Because of the directionality already present on fixed wireless links, it is likely that the gains would be less than those for the mobile case. Deployment, however, is simpler than in the mobile case as there is little need to track subscribers. In the mobile case, adaptive antennas tend to enhance the uplink rather than the downlink—for fixed wireless the most constrained link is typically the downlink because of the asymmetry of usage, so different techniques will need to be used. Another point of note is that adaptive antennas will be simpler for TDMA transmission where one array can be steered to multiple subscribers, one at a time, rather than CDMA transmissions where multiple arrays would be required to steer the different codes comprising a single carrier to different subscribers.

It would appear that adaptive antennas would be most usefully deployed if they could illuminate a subscriber using a narrow beam, but substantial work is required to understand whether the capacity gains that this would bring would be offset by the additional cost of the solution.

A further point of interest is with multiple-input multiple-output (MIMO) antennas. These use arrays at both the transmitter and the receiver; they send a different signal through each of the transmitter array antennas and then use adaptive matrix interference cancellation at the receiver to resolve the individual paths. The success of MIMO antennas depends on the path, for example, from antenna element A at the base station to antenna element A at the subscriber unit being different from that from base station element A to subscriber unit element B. With sufficient differences across the matrix, this allows the transmitted signal elements to be resolved.

In a fixed environment there is often line of sight (LOS) or near LOS from the base station to the subscriber unit. In this case, there will be no significant differences in the paths received at each of the antenna elements and so the adaptive cancellation will not work. Although far from proven, it is currently unclear whether MIMO antennas will bring key gains to this environment.

3.2.3.7 The future direction of broadband fixed wireless

This section has demonstrated that fixed wireless has a significant role to play in the future of broadband communications and is being used in areas where the copper or cable infrastructure is not appropriate or by new operators that do not have access to these legacy resources. It has also demonstrated that operators can economically and technically offer broadband services to users of 10 Mbps or more provided that they have a spectrum allocation of 100 MHz or more. Finally, it has demonstrated that there is a plethora of technical options that can be used to provide fixed wireless solutions. Of these options, it currently appears that the forerunners are the following:

- TDD/TDMA solutions;

- OFDM solutions, especially in difficult channel conditions;

- CDMA for bandwidths below 2 Mbps per user;

- Mesh for high-bandwidth, high-density systems;

- Possibly adaptive antennas, depending on future research.

Over the next few years we can expect to see a number of systems using each of these different technologies being deployed. It seems likely that the cutting edge of technology in wireless will become fixed wireless and not mobile and that this will be where many of the concepts for 4G mobile are first tried and tested.

3.2.4   Wireless LANs [5]

3.2.4.1   An overview of in-building communications

Chapter 2 discusses how the future of wireless communications might include many forms of integrated communications within and around the home or the office environment. Wireless communications within these environments could be provided in a number of different ways. The key current contenders are described as follows.

- Cellular systems, either using coverage achieved from external base stations or from special picocellular base stations placed within the building;

- Cordless phone systems designed for in-building voice communications but often with some data capabilities;

- Wireless LAN (W-LAN) systems designed primarily for data communications within the building but able to support some voice services as another form of data transmission;

- Ad hoc or personal area networking (PAN) devices that typically cannot support well-integrated communications but can link local devices together.

That there are so many different possible approaches to in-building communications reflects the fact that none of them forms a perfect solution. It is worth exploring this issue in a little more detail because it provides insight into some of the key economic and technical constraints that currently affect our capability to realize the future.

Broadly, cellular and cordless solutions come from a "voice legacy." That is, their original design criterion was the capability to carry voice traffic. To do this they needed protocols to handle mobility, they needed to be spectrally efficient (especially cellular systems), and they needed to be able to hit appropriate cost points. They were designed to integrate to other voice networks, in particular the PSTN. Wireless LAN solutions and PAN solutions come from a "data legacy." That is, they were designed to transfer data between machines. To do this they were designed with higher data rates than cellular solutions but utilized packet transmission with variable delay and quality of service (QoS) to gain the maximum efficiencies from the channel. They were not designed to integrate into many other networks, perhaps a wired office LAN but usually little more. A description of more specific issues with each of the solutions follows.

Cellular Solutions

> • Cellular solutions work in the cellular radio spectrum bands, which are typically owned by the cellular operators. Hence, in-building systems can only be deployed with the approval of the cellular operator. Worse, their spectrum utilization needs to be coordinated with the cellular network. The implication of this is that these networks probably will be owned and deployed by the cellular operators. This raises many complex issues and problems. Typically, companies own their own internal telephone systems. This provides them with control over factors such as the availability and reliability of the system and the speed at which new connections and features can be added. Ownership also enables them to provide internal calls for "free"[3] and to negotiate discount rates for external calls with long-distance and international call providers. Whether or not self-ownership is sensible, it is the manner in which most companies handle their

---

3. Of course, no call is free. It is true that there is no marginal cost for the handling of another internal call (assuming that there is capacity available), but companies will have paid for the capital cost and maintenance of their internal private branch exchange (PBX), a cost that they would not have paid were the telecommunications services provided by an external operator. Hence, the true cost of an internal call is related to the division of these costs by the number of internal calls.

telecommunications services. Since the alternative is not obviously less expensive, it takes some time and effort to explain the issues in order to change the behavior of telecommunications managers.[4] There are some possible solutions to this apparent problem. Already, in public areas, third parties are installing cellular equipment and then allowing cellular operators to interconnect their networks to this equipment for a certain fee. The building owner may be more comfortable in dealing with this third party who can get the best deal from the available operators rather than with a single individual operator in a monopoly position. (Of course, the third-party supplier is now in a monopoly position, but a rather less threatening one than the cellular operator.) Then, from the cellular operator's point of view, entering the in-building marketplace is not an obvious move. In-building networks are relatively expensive because of the need for multiple base stations in the same building compared to wide area networks where one base station covers many buildings. Furthermore, the revenue from the in-building traffic is lower on a per-minute basis than for standard cellular calls because users will expect rates inside the building to be more akin to the low rates they currently pay for fixed calls. Finally, the whole area requires intense resources from the cellular operator to negotiate a deal with the company, design and install the in-building network, assign radio frequencies, monitor the network's performance, and handle other aspects of customer care. In their present expansionary phase, cellular operators tend to be resource constrained and are thus wary of taking on this level of responsibility.

▸ Because cellular systems currently have relatively low data rates, they are not suitable for LAN applications. Although this is improving—third-generation systems are able to provide up to around 2 Mbps in indoor environments—this will take some time to arrive by which point W-LAN solutions will probably be able to offer data rates and orders of magnitude higher.

---

4. Indeed, one of the key problems here is that if the telecommunications service is outsourced, then the telecommunications managers themselves have less to do. For a telecommunications manager to recommend such a service would be akin to a turkey voting for Thanksgiving!

Cordless Solutions

- ‣ Cordless solutions also have a spectrum problem (indeed, almost all wireless communications systems do) but a rather different one from cellular systems. Cordless solutions almost invariably utilize unlicensed spectrum.[5] The key problem here is that a company has no control over the use of this spectrum. A communications system that works one day may not work the following day when a different company in a nearby building turns on their unlicensed communication system. The possibility of this happening is increasing as the number of devices proposed for unlicensed spectrum also increases. To date, there have been few reported problems because most transmissions are relatively low power and the building fabric itself provides a reasonable barrier to radio waves, tending to isolate in-building networks from each other.

- ‣ Cordless systems are different from cellular systems. This means that cellular phones will not work on cordless systems. A number of initiatives have been proposed to make "dual-mode" phones with both cellular and cordless components embedded within them, and Ericsson even produced a commercial GSM/digital enhanced cordless telephone (DECT) phone, but broadly this initiative has failed since the demand for this type of solution has been limited. (The GSM/DECT initiative is examined in detail in Section 6.5.) For users, this is a serious problem. They need different mobile phones for the office and for outside the office, and it can be difficult to provide integrated services such as a single number across these two different solutions.

W-LANs

- ‣ W-LANs tend also to use unlicensed spectrum and so suffer from the same spectrum problem as cordless solutions.

- ‣ W-LANs were not designed for voice transmission. Although voice can be carried as another form of data, typically the quality is poor,

---

5. Unlicensed spectrum is radio spectrum where any user can transmit without a need for a license or specific spectrum allocation from the spectrum regulatory body, although there are normally guidelines for the maximum power and sometimes the form of the transmission.

especially when the network is congested, because there is little QoS guarantee and so voice packets can be delayed.

▶ W-LANs do not readily integrate into telephone systems since they are designed to connect to data networks. Complex gateways are required, and W-LANs may not support the signaling needed for voice calls, for example, to carry the dialed number to the external network.

▶ Because W-LANs are different from all voice phone systems, there are no mobiles that can work on both cellular and W-LANs. Again, dual-mode devices could be envisioned, but as with cordless, there are likely to be commercial barriers to their deployment.

PANs

▶ PANs also use unlicensed spectrum and so suffer from the same problems as cordless solutions.

▶ PANs were not really designed to integrate into anything and thus can only carry voice transmissions between the devices in the ad hoc network. They may not be able to find a point of interconnection into a network that carries the signal eventually to the PSTN.

The problem that we now face is that what we really want from these in-building networks is the ability to do everything and to be integrated in a completely seamless manner. This could be achieved by a new or enhanced system or by a multiplicity of some or all of the above solutions. How it might be achieved and where these solutions might go are the subjects of Part IV of this book. The remainder of this section examines where W-LANs are today in more detail.

3.2.4.2 The current state of W-LAN technology

The world of W-LANs has been one where standardization has been successful. Almost all W-LAN solutions conform to the IEEE 802.11 standard, although the HiperLAN and HiperLAN2 standards are also contenders in this marketplace. An overview of each of these standards is provided below.

Overview of 802.11.   An 802.11 W-LAN is typically comprised of a number of base stations, or access points, in a cellular configuration. It can operate without a base station in the ad hoc fashion that PANs utilize (see Section 3.2.5) but this mode of operation is far less common than the cellular mode. The IEEE 802.11 protocol stack consists of the lower physical layers and then a common layer 3 protocol. There are three different physical layers defined, listed as follows:

▸ Frequency hopping in the 2.4-GHz unlicensed band;

▸ Direct sequence spread spectrum in the same band;

▸ Infrared.

Each of these offers data rates of 1 or 2 Mbps depending on the configuration. In 2000 there were plans for a number of different versions of IEEE 802.11 including an 11-Mbps version known as 802.11b and a 54-Mbps version known as 802.11a. These required a modification of the FCC rules for use of the unlicensed frequency band, and during 2000 it was not clear whether this change of rule would be forthcoming or when it might happen. The 802.11a standard is very similar to the HiperLAN2 standard, and it is possible that these will merge into one single standard.

The basic multiple access method used is carrier sense multiple access with collision avoidance (CSMA/CA). Essentially, a device that wishes to transmit a packet monitors the radio channel. If it detects activity, it does not transmit. If it senses no activity, it waits a random time before sending its packet. It is possible that another device also transmitted its packet at the same time. To overcome this, the base station sends an acknowledgment for each packet received. If the transmitting station does not get an acknowledgement, it waits a random length of time before retransmitting its packet. In the case where there are relatively long data frames to send, the device can send a request to send (RTS), a short packet containing only its identity and the length of the information that it wishes to send. If this is received correctly, the base station will then allocate some resource, mark the channel as busy to prevent other devices from accessing, and then the full message can be sent.

Overview of HiperLAN2. The basic architecture of HiperLAN2 is very similar to that of IEEE 802.11 in that there are a number of base stations, and devices can either communicate with the base station or directly with each other. The base stations, or access points, can automatically configure their frequency so that there is no need for manual frequency assignment.

HiperLAN2 offers the potential of much higher data rate transmission than 802.11. Physical transmission can take place at up to 54 Mbps, resulting in a user data rate after error correction and other overheads of 25 Mbps. The air interface used for this is OFDM, which segments the incoming data stream into a number of subsidiary streams and transmits them on subchannels. This avoids some of the problems associated with intersymbol interference when wideband transmissions are utilized. HiperLAN2 is connection-oriented. This is similar to the RTS mode in 802.11, but in HiperLAN2 all transmissions must start with an RTS that includes the bandwidth required. The bandwidth is then reserved by the access point before the transmissions take place. This feature also allows a certain amount of differing QoSs to be supported. HiperLAN2 supports mobility and allows the users to roam from one access point to another; it also allows message flows to be handed off as the devices move across the boundary point between the cells.

There are three main layers in the HiperLAN2 protocol stack: the physical layer, the data link control layer, and the convergence layer. These are described as follows:

> *Physical layer.* This is comprised of 20-MHz wide channels in the frequency bands 5.15–5.35 GHz and 5.47–5.725 GHz. This is a total of 455 MHz in Europe (less is other parts of the world), although the upper band is unlicensed so there may be interference from other users of the radio spectrum in this band. OFDM is deployed and divides the channel into 52 subcarriers of which 48 are used for data; the remaining four are pilots used to monitor the phase change in the channel. Each packet contains a guard band to allow for multipath reflections of 800 ns (typically the delay spread in the indoor environment does not exceed 250 ns). Several modulation levels and coding rates are defined in the standard, ranging from binary phase shift keying (BPSK) to 64 QAM and coding rates of 1/2 to 3/4. This leads to a physical layer data rate of 6–54 Mbps depending on the environment and the capabilities of the devices. The framing is based on TDD/TDMA with dynamic scheduling between uplink and downlink (that is, the relative size of these links can change if the data is asymmetrical). Each frame has a basic duration of 2 ms.

> *Data link control layer.* This provides the medium access control (MAC) protocol, the error control protocol, and the radio link

control protocol. It involves the provision of a number of control channels, such as a broadcast channel, which provides network information, and a frame control channel, which details the allocation of the resources. There is also a random access channel to enable devices to request resource.

> *Convergence layer.* This acts mainly as a buffer between higher and lower layers, adapting service requests received from the higher layers into the appropriate format and framing size for the lower layers. Because there is a fixed frame size at the lower layers but variable-sized packets may arrive at the higher level, segmentation is a key function of this layer.

The HiperLAN2 standard suggests a number of possible applications for HiperLAN2, including the following:

> *Corporate LAN.* This is the classical W-LAN deployment whereby access points are connected to the network backbone, and devices communicate to these access points using HiperLAN2.

> *Hot spot.* This would be the use of HiperLAN2 in places such as airports. Devices would access the airport network, and then tunneling, or something similar, could be used to connect back to the corporate network.

> *Access to third-generation cellular networks.* The suggestion here is that high-density areas would be covered with HiperLAN2 while lower density areas would be covered with third-generation cellular solutions. Given that HiperLAN2 is not well suited to voice transmission, this seems somewhat unlikely at present.

> *Home.* This is the use of HiperLAN2 to provide a home network into which devices such as computers and video cameras can be interconnected. Although plausible, it seems more likely that BlueTooth (see Section 3.2.5) would play this role because of the much lower cost of a BlueTooth chip compared to a HiperLAN2 chip.

Table 3.3 provides a comparison between HiperLAN2 and the variants of IEEE 802.11.

Table 3.3
A Comparison Between Different W-LAN Technologies

Characteristic	802.11	802.11b	802.11a	HiperLAN2
Spectrum	2.4 GHz	2.4 GHz	5 GHz	5 GHz
Max physical data rate	2 Mbps	11 Mbps	54 Mbps	54 Mbps
Max user data rate	1.2 Mbps	5 Mbps	32 Mbps	32 Mbps
MAC	CSMA/CA	CSMA/CA	To be determined	TDD/TDMA
Connection	Connectionless	Connectionless	Connectionless	Connection-oriented
Frequency selection	Frequency hopped (FH) or direct sequence spread spectrum (DSSS)	DSSS	OFDM	OFDM with dynamic frequency selection

W-LAN commercialization. Although the standardization may have been successful, commercialization of W-LANs has been less so. Initially, the W-LAN proposition was that the wired office LAN could be replaced with a wireless LAN that would enable computers to be moved around the office without needing to modify the wiring. Proponents postulated that companies spent substantial money on network management as individuals moved location or position or left and joined the company. W-LANs also allow people to use computers away from their desk, in meeting rooms or other locations. Very few companies accepted this argument. They found that fixed LANs were evolving at a rapid pace and provided a capacity and maximum data rate much higher than that offered by wireless LANs and were developing tools to simplify the management of change. Companies were also concerned about the cost and reliability of W-LANs, and, at least initially, the prices were substantially higher than for traditional wired solutions.

This initial failure does not appear to have unduly affected the W-LAN industry. There are many innovative companies emerging with new technologies and solutions and innovative ideas for their use. For example, one company has developed a W-LAN system that can transmit high-quality video to handheld terminals. The plan is to lease these terminals to spectators at sporting events and locally broadcast to the terminals immediate replays of critical events in the game, allowing spectators to have both the live atmosphere and the multicamera replay capability that home viewers enjoy. Few of these companies to date, though, appear to be addressing the convergence of voice and data solutions into a single integrated in-building communications system. Most are targeting data rates in the region of 5–10 Mbps and all are proposing the use of unlicensed spectrum.

3.2.5 Wireless PANs [6]

3.2.5.1 What is a PAN?

The concept of a personal area network is that an individual may have a number of devices that they carry on or near their person. These might include a mobile phone, a PDA, a laptop computer, and a personal stereo. There are many potential advantages of these devices being able to communicate to each other, especially without wires. For example, today it is possible to link the laptop to a mobile phone through a cable connector and to wirelessly link into a remote data network to retrieve e-mails or perform other actions. This would require that the individual carry the necessary connectors, have the space to lay out the laptop and phone, and understand the often complex actions required to make a connection between the different devices. The concept of PANs is that if each of these devices had a short-range communications tool built into them, they could exchange information without wires and without any intervention from the user. So, for example, the laptop, which was ostensibly in "sleep mode," stored in a briefcase, could periodically talk to the cell phone clipped to the user's belt and ask it to check for e-mails. The cellular phone could retrieve these and send them to the laptop over the short-range link. The laptop could then store them so that when the user turned the computer "on," all the e-mails would be available on the computer. As the user performs actions such as sending e-mails, the computer would talk with the mobile phone and request transmission of these to the data network.

This concept, which is at the heart of the BlueTooth standard discussed in Section 3.2.5.2, results in a number of unusual requirements and properties, often talked about under the term "ad hoc networking."

▸ The short-range devices must be very inexpensive because they need to be embedded in most machines. Price points of a maximum of $5 per chip are often discussed.

‣ This price requirement in itself tends to limit the devices to a very short range since one of the major cost elements of a wireless device is the RF amplification, with prices rising dramatically as power levels rise.

‣ The system needs to utilize either dedicated or unlicensed spectrum and needs to be "self-organizing" from a spectral point of view since there can be no central control as in cellular systems to organize spectral reuse. In practice, there is unlikely to be dedicated spectrum for these kinds of devices, so they are designed to use unlicensed spectrum.

‣ Networks need to be "self-discovering" and "self-managing." No one device can know beforehand what other devices might be in the vicinity, so devices need to have a method for discovering other devices, finding out what they are, and transmitting appropriate information to them.

‣ With an almost infinite range of devices, understanding how devices communicate, what information they might want to send to each other, and whether they have appropriate rights to do so becomes a major problem.

PANs as they were originally conceived were not designed to handle voice transmission. They were for machine-to-machine communications rather than communications involving people. Many have realized, however, that if PAN communication chips become ubiquitous, then they could form the ideal basis for local voice networks as well. This is something that we will return to when we look at the future in Part IV. Suffice it to say here that, like W-LANs, PANs are generally not well designed for voice transmission as they tend to delay packets when the network becomes congested. However, in a PAN environment there may be limited congestion and so this may not be a key issue.

3.2.5.2 The BlueTooth standard

BlueTooth is a PAN standard providing short-range ad hoc communications between a number of disparate devices. It is designed to operate in the unlicensed industrial, scientific, and medical (ISM) band centered around 2.45 GHz. The bandwidth of transmission is 1 MHz, but the actual 1-MHz band utilized within the 80-MHz-wide ISM band can be selected by the device, depending on the environment. In order to overcome the anticipated interference within the ISM band, BlueTooth utilizes frequency hopping such that a single (static) source of interference will only have a limited effect on transmission quality. Within the 1-MHz bandwidth a data rate of 700 Kbps is used. This is on a TDMA framing structure with different parts of the frame set aside for voice and data, allowing different quality of service between the different services. During 2000 there was also discussion about a BlueTooth 2.0 standard, which would enable data transmission at rates of around 10 Mbps and would probably become available around 2003.

Two levels of transmit power are available within the BlueTooth specification: 1 mW (0 dBm) and 100 mW (20 dBm). The former allows transmission ranges of up to around 10m, the latter up to around 100m. There is still some debate over which of these power levels is more suitable. The lower power levels result in inexpensive devices and minimal interference to other neighboring PANs. The higher power level is more expensive to implement and has a greater drain on the battery but enables simpler coverage of a house or other building. The higher power devices utilize power control to enable them to communicate with lower power devices and to reduce transmission power where full power is not required.

One of the largest problems for BlueTooth is detecting the presence of other users and building an ad hoc network. This is exacerbated by the need for BlueTooth-equipped devices to generally have low battery drain. The only way for one BlueTooth device to discover another is for at least one of them to transmit periodically; however, the frequency of transmission is kept low to minimize battery usage. Since the device will only transmit for a short time period and may do so on any one of the numerous different frequencies in use, the receiver must listen for as long as possible. This also drains batteries. Complex algorithms are in place to enable devices to become aware of each other with a minimal battery drain but within a sufficiently short time period that the user is not waiting for the devices.

Once connected, there are additional difficulties associated with building and running ad hoc networks. These concerns are outside the scope of this book. Suffice it to say that solutions to these problems have been found, and BlueTooth systems look likely to work together well.

Table 3.4 shows current predictions of the size of the BlueTooth market. This only considers the most obvious segments for BlueTooth usage and does not look at areas such as household appliances. Nevertheless, by 2005 it is predicted that sales of BlueTooth-equipped units will exceed

Table 3.4
Volume of BlueTooth-Equipped Devices Being Sold (in Millions)

Units sold (m) per sector	2000	2001	2002	2003	2004	2005
Mobile terminals	3.9	25.4	83.7	251	588	933
Mobile computing	0.7	4.5	13	23	39	60
Desktop computing	0.4	4	12	24	40	63
Set-top boxes	0.1	1.2	2.5	5	10	18
Automotive	0	0	0.7	2.2	3.8	7
TOTAL	5	35	112	305	682	1,083

Source: IMS, 2000.

1 billion units per year. It might be expected that by 2010, most people in developed countries will own multiple devices equipped with BlueTooth-like communication systems.

3.2.6 Ad hoc devices

This group contains the many communication devices that each serves one particular purpose—garage door opener, television remote control, keyless entry system for the car, remote control toy, and many, many more. Simply describing these as ad hoc gives some good guidance as to the characteristics of the systems in this group. Typically, devices in this category utilize unlicensed spectrum. They are typically short-range devices. They do not conform to a particular protocol or type of message transmission, although television remote controls as a whole do tend to be interoperable to some degree.

3.2.7 Networks [7]

Networks are not wireless communications devices—typically they are wired—but as will become apparent, many constraints on the future, especially on the interworking of different wireless communication devices, are imposed by the networks rather than the wireless devices. Hence, an understanding of networks themselves is important to understand the future.

For many years there was only one network—the PSTN. Historically, this has been a network comprising copper twisted pair wiring to the premise connected back to a number of central switches, possibly through subsidiary switching nodes that act to concentrate the connections. The role of the central switch was to make a "circuit-switched" connection[6] between any two premises connected to the network. The form of the connection was an analog channel with a bandwidth of around 3 kHz and a signal-to-noise ratio (SNR) of 30–40 dB. This was fine for voice and could be utilized for data transmission by adding modems to the data devices, which transformed the digital information to be transmitted into an analog waveform that maximized the capacity of the channel. More details about this form of data transmission can be found in [8].

The PSTN has slowly been evolving. Switches have become more intelligent, and they're now able to provide supplementary services such as call forwarding. Backbone networks have become high-capacity digital systems so that information is carried from one exchange to another in a digital form. An overview of switching and fixed network transmission can be found in [3, 7]. These networks, however, remain circuit-switched and have much heritage and legacy support, which means that any change must be slow.

Another global network has emerged in the last decade—the Internet. It can be difficult to differentiate the Internet from the PSTN as they use many shared components. A typical residential user dials into the Internet using the PSTN. The PSTN makes a circuit-switched call between the residence and the Internet service provider (ISP), which, as far as the PSTN is concerned, is just like any other call. Modems owned by the user and the ISP enable computers to communicate over this link. Software in these computers enables Internet commands to be sent. When the residential user "clicks" on a particular page, the ISP receives this request and sends it into the Internet. Broadly, at this point, the Internet is a number of ISPs and other nodes, all connected together in an ad hoc fashion. The ISP will consult routing tables to establish where the request for information should be sent and then utilize broadband digital links to retrieve this information. These may be the same broadband links that the PSTN uses to connect switches, but typically different "virtual" connections across this medium will be utilized to give the impression of each user having their own dedicated link.

Although they share many components, the PSTN and the Internet are logically different networks. The PSTN would not understand an Internet address, and the Internet would not know how to treat a phone number. The PSTN cannot handle Internet data, but the Internet is slowly

---

6. That is, a connection that remains in place for the duration of the call providing a dedicated path between the two parties in the call, which is present even if neither of the two parties transmits information.

acquiring the capability to provide voice communications. This lack of integration is a problem for the provision of unified services—the delivery of all messages whether voice, e-mail, or fax, to individuals, wherever they are.

Next there are cellular networks. A typical cellular network is shown in Figure 3.3. It comprises a number of base stations providing coverage to cellular subscribers. These may be linked to base station controllers, providing a concentration function, which, in turn are linked to cellular switches often known as mobile switching centers. Cellular switches are similar to PSTN switches but with enhanced functionality to handle the mobility inherent in cellular communications. If a call from a cellular subscriber is destined for the PSTN, the cellular network will utilize a gateway node to deal with any protocol differences and make the physical connection. Key to the debate that will form much of the latter parts of this book is that cellular switches and PSTN switches are different, they use different protocols, they provide different services, and they are

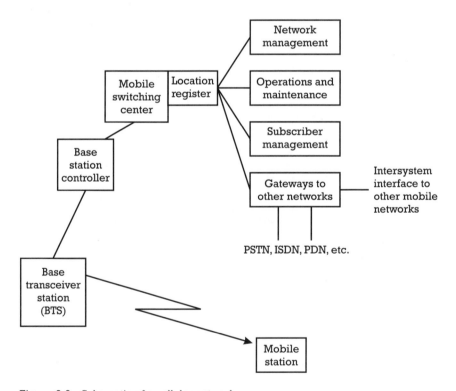

Figure 3.3 Schematic of a cellular network.

difficult to interconnect in any fashion more complex than sending and receiving call information.

Fixed wireless networks come in many varieties. Some are focused on voice provision and form a hybrid between cellular and PSTN networks. They have base stations and base station controllers but connect directly to a PSTN switch. This is possible because there is no mobility in a fixed network (by definition). Even so, there are often difficulties in simple areas such as knowing when the subscriber has entered all the dialed digits. In a cellular system, the subscriber composes the number and then presses "send." In a PSTN system the subscriber simply composes the number—the network recognizes when it is complete. Fixed wireless systems, being somewhat of a hybrid, need special arrangements to overcome these issues. Other fixed wireless networks that do not carry voice (or at least, treat voice just as another form of data) may have no connection to the PSTN but instead connect directly to the Internet, or an operator's data network, which in turn has an Internet gateway. Still others may provide only a dedicated connection between two points and as such look, to the PSTN or Internet, just like another link between two points, to be used as a physical resource for transmission of any sort of information.

In addition to these global networks, there are millions of smaller networks, typically termed LANs. Some of these reside within a building. Others, say, for multinational corporations, can be global, linking all their different sites together. Again, these may share infrastructure; for example, a global corporation will typically not install their own cable to link disparate sites but will lease capacity from companies specializing in the provision of long distance links (often the PTOs). This may be the very same cable used to link PSTN switches and to provide Internet connectivity, but as before, the use of virtual provisioning will make it appear to the corporation that they have a dedicated connection.

3.2.7.1 Addressing—the heart of the problem

As stand-alone networks there is not a lot that is wrong about these individual networks. The key problem is to make them work together in an integrated fashion so that users can have fully integrated communications such as the capability for all calls to reach them wherever they are. Understanding how addressing work illustrates the difficulty in the problem. When a phone number is dialed, such as 847–123–4567, say, from a fixed phone, the PSTN takes the first group of digits, 847, and uses these to route the call to the local exchange responsible for the 847 numbering set. This exchange has many of the individual lines under the 847 area code connected to it and can make the connection with the right line. Now imagine calling a cellular number, say, 847–309–1234. The call is routed in the same manner until it reaches the local exchange, which understands that this call is not for a number set under its control. It consults a look-up table and determines that the call should be forwarded to a cellular switch. This takes the number and passes it to the home location register (HLR), a database that keeps track of where each subscriber is. Based on the information it receives, it forwards the call to the mobile switching center (MSC) that controls the cells where the subscriber is currently located. This MSC then pages the subscriber in order to start the process of making a connection.

Now consider the case of an e-mail address—john.smith@cellular.com. This is passed into the Internet and sent to one of a number of sites that contain domain routing information. This is effectively a large look-up table that can translate the e-mail address into a numerical address of the form 123.202.101.200, each number being less than 256 since it is represented in binary form by eight binary digits. The first one or two numbers are then used to determine a routing path to the geographical location relating to that particular number. The last two sets of digits are typically used within the organization to route the e-mail to the correct user.

What if the user would like all calls to come to his mobile when it is turned on? If his mobile number is called, then the call will automatically go to his mobile, although if it is turned off, it will typically go to his mobile mailbox (some mobile systems allow for calls to be forwarded to a predesignated number if the mobile is turned off). If the fixed-line number is called, there is currently no way for the fixed phone system to know that it needs to check to see whether the mobile phone is on. Indeed, the whole concept of "on" is not understood by fixed networks since fixed phones are always on, simply either busy or not busy. The only possibility is to set up a way to forward all calls to the mobile phone and then have the mobile phone forward calls back to the fixed phone in the case that it is not turned on. Unfortunately, many fixed phones do not offer call-forwarding services, so this will not always be possible. The individual also wants to receive e-mails on his phone since he has a large display and can read them. Because of the addressing system, the e-mails will always go to his office computer first. It would be possible to write software within the office network that forwards the e-mail; however, the office network is unable to determine whether the individual's mobile is on since it has no access to the HLR or any common protocol that it could use to communicate with the HLR. One way around this problem would be that when the mobile phone was turned on it sent a message automatically to the home LAN telling it that it was now turned on and wanted to be sent e-mails. There are further difficulties in how to use Internet protocol (IP) addresses to send a message to a mobile device, although these are now being solved by the mobile IP group. Upon turning off the phone, the phone would need to send another message to the home network telling it not to forward any more messages.

This complex set of actions, which would not work in all cases, shows that many of the things that an individual might want to do are just about possible in a roundabout and complex manner. They are made complex by the lack of communications between different networks and the rigidity of the addressing rules. As will be discussed in Part IV, what is really required is a location-independent addressing scheme and widespread access to databases containing aspects such as location and routing information for individual subscribers.

3.2.7.2   Packet protocols, IP, and asynchronous transfer mode (ATM)

It is worth talking briefly about packet data protocols. This is because, as Section 3.2.7.3 shows, it is almost universally accepted that the telecommunications networks of the future will be packet-based networks utilizing packet protocols for the transmission of information. A short description of these protocols provides the necessary background information to understand the limitations of these kinds of networks. For a more complete tutorial on packet data protocols, see [3].

There are many protocols that are utilized for packet data transmission. These include X25, frame relay, IP, and ATM. Of these, IP and ATM are considered to be the two key contenders for future networks. All packet protocols work by taking the incoming information stream and dividing it into a number of sections, each section being of a particular length. They then encapsulate this information into a packet by adding header information that describes aspects such as the destination of the packet and identifies the transmitter. They may also add error protection to enable errors in transmission to be detected.

IP was originally designed for the transmission of non-time-critical information across a potentially hostile network. It is non-connection-oriented. That is, if a message needs to be split into a number of packets, it sends each packet into the network independently, without asking the network to establish a particular channel so that all messages follow the same path. As a result, it is entirely possible for a packet sent third to be received second and the packet sent second to come along some time later. The framing format includes the header length, the type of service, the total length, the identification, flags, a field describing how long the message should be held before being deleted in the case of a problematic delivery, protocol details, a header checksum, source and destination addresses, options padding, and finally the user data.

IP is optimized for the Internet (hence the name). It is less well suited to voice transmission. In voice transmission, delay is critical. If the delay exceeds around 100 ms, then it will become noticeable and disturbing to humans. Voice is also a relatively low data rate (compared to many Internet transactions). It is typical to place around 20 ms of speech in one packet. This is because it will take 20 ms to accumulate the speech to put into the packet before it can be transmitted, so the longer the packet, the longer the delay imposed at the transmitter. Because the network and the receiver will add additional delay, 20 ms is typically all that can be afforded at the transmitter end. It is common to encode speech at a data rate of around 8 Kbps, depending on the quality required. At 8 Kbps, 20 ms of speech is 160 bits, or 20 bytes. This is a very small load for an IP packet, where the header information amounts to some 32 bytes, depending on the information coded, hence adding redundancy of over 100%. This is one problem with IP. Another is that it is difficult to mark speech packets as important, so a speech packet may wait in a queue while a very long data packet is sent, adding critical delay. Some realizations of IP, such as proprietary versions created by Cisco, do allow QoS parameters to be added to the message, but the availability of these cannot yet be guaranteed across the Internet.

ATM is optimized for dedicated telecommunications networks where the network is assumed to be reliable and "friendly." ATM is connection-oriented, that is, at the start of a transmission process, a virtual path is established across the network. All the packets belonging to this particular transmission process then follow the same path. This requires some time to set up the connection, but after this has been done, packet headers can be very small, containing only details of the virtual path and not needing to hold the destination and origination address. Unlike the other packet protocols discussed, ATM uses packets of a fixed size. These contain a 5-byte header and 48 bytes of user data.

The use of a small fixed size ensures that if a high-priority packet arrives at a router at the point in time that the router has just started sending a low-priority packet from a different user, then that low-priority packet will only take a short length of time as it is only a small size.

Hence, delay can be kept to strictly controlled limits. ATM does have QoS parameters, many of them relating to how the packet should be treated and how excess bandwidth should be distributed if it is available.

Both protocols have advantages as they are both optimized for the particular networks for which they were originally designed, and some have postulated using both into the future. Using a multiplicity of protocols, however, has disadvantages. There is the need for protocol conversion, and, for example, if ATM is converted to IP for part of its transmission, all the properties of ATM such as QoS may be lost, rendering the use of ATM irrelevant. Many have argued that the world needs to standardize a single protocol for data transmission and that, because of the ubiquity of IP compared to ATM (most home computers talk in IP, whereas only specialized telecommunications nodes talk in ATM), IP should be the protocol of choice. This is a view strongly held by Cisco, the world's leading manufacturer of data network equipment, and as a result, is likely to come to pass. Through various standards associations and through proprietary developments, companies like Cisco are likely to slowly overcome the limitations of IP to the point where it is the protocol of choice for all future data transmissions.

3.2.7.3 The all-IP vision

It is generally believed that the network of the future will be packet-based [9]. The reason for this is that packet networks can carry both voice and data traffic efficiently across a single network, whereas circuit-switched networks, such as the PSTN and today's mobile networks, can only transport data on dedicated connections. Dedicated connections are highly inflexible: If the data rate is lower than the connection capability, then the excess capacity is wasted; if the data rate is higher than the allocated bandwidth, it cannot be carried. The problems with packet networks tend to relate to delays and reliability, but these are slowly being overcome. The generally agreed vision [9] of how the future network will look is shown in Figure 3.4.

Some of the key points to note with this network vision are described as follows:

> • One core network serves a multiplicity of different access networks. This is in contrast to the situation today where each different access network has its own specialized switching network (e.g., the MSC in mobile networks). This brings a number of advantages both in

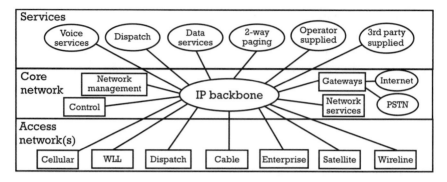

Figure 3.4   The generally agreed vision of a future network.

economies through the reuse of a single network and the ability to provide the same services across multiple access networks because they all use the same common core. Here, a range of access networks—including cellular, wireless local loop (WLL), dispatch systems such as the iDEN technology, cable-based networks, systems for companies or enterprises, satellite, and conventional wireline networks—are shown connected to the same core.

▸ The central part of the network is an IP backbone. This consists of a number of routers with associated routing tables. There is very little intelligence in this part of the network, simply the capability to transmit information from any incoming point to any outgoing point. This is in contrast with conventional networks where all the intelligence resides within the switch, which is also the device that performs the "routing." The conventional approach makes it difficult to add new services or modify existing services because it is typically only possible for the switch manufacturer to modify the service scripts within the switch.

▸ Surrounding the IP backbone part of the network is a number of "housekeeping elements," that is, elements that are required for the correct functioning of the core network. These include the network management, which monitors the correct operation of each node, the control, which allocates bandwidth between different routers, the gateways, which provide protocol translation to allow interconnection to external networks such as the Internet or PSTN, and the network services, which provide core services that cannot be separated from the network.

• There is a service plane where the service scripts reside. These may be as fundamental as the scripts that control the manner in which a voice call is made, running voice call models such as those in the H.323 standard. Key to the concept is the provision of third-party servers, which can provide services to the core network without the intervention of the manufacturer of the network equipment, and perhaps even without the intervention of the operator. If a user desires to use one of these new services, all that is required is that when they initiate the "call" they provide the address of the server where the service script is stored. This address may be supplied to the user terminal when the user responds to an advertisement or after browsing on a particular Internet site.

It is clear that such a network brings key advantages to the operator, as already discussed and summarized in the following list:

• Reduced cost of operations since only one network needs to be maintained for all different access networks and for both voice and data communications;

• The ability to carry multimedia traffic (that is, traffic that contains multiple types of information such as voice and text information combined);

• Simple access to Internet-based applications through the use of a common IP protocol and addressing mechanisms;

• Rapid provision of services through the capability for multiple parties to add service scripts independent of the underlying network architecture. This is akin to the ability of multiple third parties to develop software for computers using a common Windows operating system.

Key among these benefits, as far as the view of the future is concerned, is the capability for multiple different network types to connect to one common core backbone using a common IP protocol. Using such a network, the example of calls being forwarded to an individual's cellular phone if it is turned on becomes much simpler. What happens in this new model is that whatever source the call arrives from, it is handled by the voice services client in the services layer. This client provides a consistent service regardless of the source of the call. The client can be programmed with the knowledge that it should forward the call to the mobile if the mobile is turned on. The service client can then access a common location register, available to all entities in the service plane to determine whether the mobile is on or not. If it is, then the service client can instruct the call to be established to the mobile phone. It is the removal of the intelligence from the access network to a common service plane available to all the different networks that provides the capability to simply deliver services of this sort on future networks.

3.2.8 Broadcasting

Broadcasting does not currently sit within the definition of wireless communications, primarily because it is both one-way and one-to-many communication. Nevertheless, there may be sufficient convergence in the future to make an understanding of the world of broadcasting worthwhile.

Broadcasting broadly covers audio and video broadcasting—both can be treated in the same way for the purposes of this discussion. In the case of video broadcasting there are the following mechanisms:

- *Terrestrial broadcasting.* Typically uses UHF channels to broadcast video signals. This remains the predominant means of broadcasting in most parts of the world.

- *Cable broadcasting.* Sends multiple channels down a cable. Since this is not wireless, it is not a key focus for this book.

- *Satellite broadcasting.* Uses a geostationary satellite to broadcast multiple channels to receiver dishes mounted on the side of homes. This is a mechanism currently gaining ground because of the number of channels that can be distributed and the relatively rapid reach to the home.

- *Internet broadcasting.* This is still in its infancy but some are now developing streaming media systems whereby if end users log onto a particular site they will receive whatever "station" is currently being broadcast. Although the end user would perceive this to be broadcast, from an architectural point of view, since the signal is sent individually to each subscriber, it is more of a multiple one-to-one communication than a true broadcast.

There are a number of key questions to discuss when considering where broadcasting is likely to go in the future and whether it will affect the types of wireless communications discussed in this section.

> • *Will people want broadcast entertainment?* There is an argument that suggests that users are moving toward increased personalization. In such a scenario, users tell a broadcast application provider on the Internet the type of content in which they are interested and the amount of time they want to spend watching (or listening to) the content. The site then assembles a custom set of programming and dispatches it to the user. Thus, although many people might watch the same clip of video, they do so at different times and as part of different programming. Broadcasting as an architectural way to reach multiple subscribers simultaneously would have ceased to exist. Although apparently compelling, the argument for personalized broadcasting is somewhat weakened by the cultural change that would be required to bring it about. People enjoy gathering together to watch sporting events, or tuning in every night to watch their favorite soap opera. Cultural change of this sort takes at least a generation, and so although we might see significant change in the next 20 years, we are unlikely to see the death of broadcasting as we currently know it.

> • *Will there be sufficient bandwidth for personalized services?* One of the key problems with the delivery of personalized services is the huge amount of bandwidth it would require. Each person now needs a multimegabit connection to view video content (typically around 2 Mbps for most content). Providing this to even a minority of the users simultaneously, as will be demonstrated in Chapter 4, is beyond the capabilities of cellular networks, even those of 20 years from now. Providing it to the home, however, using fixed wireless and a high-speed fiber backbone is perfectly feasible. Furthermore, assuming that a cellular user has a portable device, he or she would probably not need high-quality video viewing and may accept video quality at a much lower bit rate. So, in summary, sufficient bandwidth may be available assuming most users want to view the content in their homes.

It would be technically possible and attractive to many to move to a world where broadcasting is personalized. However, that would require significant cultural change of a form that is likely to take a generation to affect—and whatever happens, people will still want to tune in live to the World Series in a broadcast mode.

3.3 How well does the current position match user requirements?

We ended Chapter 2 with a very broad definition of what a user might conceivably want in the next 20 years. Now that we have a better understanding of where we are today, it is appropriate to ask where we are in terms of providing these broad requirements. Today it is possible to communicate to anyone else who has a communications device assuming that it is both turned on and in coverage and that the form of communications is voice. With some systems, such as iDEN or the advanced speech call items (ASCI) [10] within GSM, it is possible to talk to a group of people, and it might be expected that such group communications will become more widespread in the next few years.

There remains the problem of individuals having multiple communications devices that are not integrated in any meaningful fashion. The obvious solution to this is for the mobile device to become the primary communications device, replacing the office phone and the home phone. It can also replace the pager since it can receive, display, and send text messages in just the same way as a pager. This raises additional problems associated with how this will be achieved in terms of the following.

> • Integration with the office and home wireless networks;

> • Intelligence in the network such that incoming messages are dealt with appropriately given the time and location of the user;

> • How the mobile device will be linked into the computer system for the forwarding of e-mails given the rather different network concepts of voice and data systems.

These are problems to which we will need to return in Section 16.3 to look at how they might be solved and what impact they will have on future communications systems.

It is not possible to communicate regardless of location—there are many locations that are problematic, including aircraft (although this is improving, it only allows people on the aircraft to call to the ground, not the other way around, and there are no initiatives in place to change this), ships, rural areas, parts of the world where different cellular standards are in use, inside buildings, and so on. This is slowly improving. Land-based coverage is continually getting better and can be expected to extend to within buildings in the coming years. Standardization of cellular technologies around the world is slowly happening, and where it is problematic, multimode phones are being developed. The coverage problem, then, will slowly disappear over the coming years for all but a very few remote locations and can be dismissed from further discussion. We will simply assume that current technologies will solve this problem.

The final aspect is the form of communications. Speech is already available and is the default medium. Text is generally straightforward, albeit sometimes difficult for the end user because of the difficulty of entering and reading text on small communications devices. This is not the whole story for text transmission to and around the home or office where files might include substantial graphics content, and we will return to this later in this section. It is video that is far from being realized in anything other than demonstrators. There are a number of aspects to the video problem, including the following:

› Integration of a camera into a mobile device;

› Integration of a suitable screen to allow the viewing of video imagery;

› Social and ergonomic issues related to the use of video communications in a portable device;

› Provision of suitable bandwidth in the wireless link to provide video communications of sufficient quality;

› Common protocols for video transmission such that the signal can be reconstructed on any machine connected to any network.

Some are more problematic than others. Small cameras and screens already exist, and it can only be a matter of time before they are integrated into portable communications devices—probably less than five years from now for widespread availability. Social and ergonomic issues can undoubtedly be overcome, and users will soon learn that video communications are only meaningful where they have the time and space to interact in an appropriate manner. Common protocols are available, using the Moving Picture Experts Group (MPEG) standards, of which there are a number, depending on the quality required. Bandwidth, however, is more problematic.

The bandwidth required depends on the quality of video communications required, which is itself dependent on these factors:

▸ The resolution required in the end picture (which, we can assume, will be relatively low for most wireless communications to mobile devices but higher for wireless communications to fixed devices);

▸ The degree of motion in the picture, about which we can assume less, but in general, if the video information is a picture of a person, the degree of motion is relatively low; however, individuals might want to point the camera at, for example, their child playing, to relay these pictures to a remote location.

The bandwidth is also dependent on the capability of the coding algorithms deployed. Table 3.5 provides some information on historical and current video coding rates and predictions that have been made on progress.

Note that where there is a range of coding rates, the video coder can support any of these rates but with varying quality, such that the encoder can modify its behavior according to the channel available. Video standards are developed by the MPEG. There are three standards:

▸ *MPEG-1:* Designed to deliver full-screen video at near VHS quality. The compression ratio can be up to 200:1.

Table 3.5
Video Coding Rates in the Last 10 Years

	Low-Quality Video	**High-Quality Video**
1991	H.261 at 64 Kbps	MPEG-1 at 1–4 Mbps
1993	—	MPEG-2 at 1–8 Mbps
1996	H.263 at 60 Kbps	—
2000	MPEG-4 SP at 58 Kbps	MPEG-4 at 1–8 Mbps

> ‣ *MPEG-2:* Designed for digital satellite, cable, and terrestrial broad-
> cast, as well as digital versatile disk (DVD); sometimes used by inter-
> active TV and delivering a quality comparable to SVHS.

> ‣ *MPEG-4:* Both a low-bandwidth format used across the Web and
> cellular and a superset of MPEG-2 high-quality images.

What is clear is that there has been little change in video coding rates over the last 10 years. Some experts expect this to continue and predict that video coding rates will not fall much in the coming 10 years unless there is some breakthrough discovery. It is very difficult to extrapo-late beyond this, but as a first pass we might predict that video coding rates will stay approximately constant over the time horizon we are considering.

From Table 3.5 it can be seen that for the sort of video communica-tions that might, at least initially, be adequate for mobile wireless com-munications, a bandwidth of around 60 Kbps would be sufficient. This bandwidth is becoming available on enhanced second-generation cellular systems and is well within the capabilities of third-generation systems. So this is technically feasible. The key issue here is the economics of the provision of this service, and we will return to this in Section 16.3.5. To the home, it is more likely that users will require higher resolution since the video call might be displayed on a large screen. Here they might opt for rates in the order of 1 Mbps if economically viable. Hence, we have another issue to return to, namely, the economics of the provision of video communications to wireless devices.

It was mentioned above that large text or multimedia files might be problematic. It is quite common today for files containing detailed graphi-cal content to exceed 10 Mbytes, and we can imagine that as more video and audio components are added to such documents that their size will grow rapidly maybe to 100 Mbytes or more. We can transmit these files today to wireless devices or wired devices connected to almost any net-work. The problem is the time taken to transmit them. Standard trans-mission rates for fixed networks to the home are on the order of 30 Kbps, at which speed a 10-Mbyte (80-Mbit) file takes some 44 minutes and a 100-Mbyte file some 7.5 hours. At the data rate common to most mobile phones (9.6 Kbps), these times increase by approximately a factor of three. The length of transmission time that the user is prepared to toler-ate will depend on the user and on the importance of the file, but clearly the times listed above are unreasonably long for most people. Ideally, people would like the files to arrive near instantaneously, which probably means within a few seconds. For example, in order to achieve a five-second download time for a 10-Mbyte file, it would require a data rate of 16 Mbps. Achieving this same download time for a 100-Mbyte file would necessitate 160 Mbps. These data rates are probably not needed for hand-held devices, which will have limited capabilities to store and process information of this sort. It will be required on fixed networks connected to homes and it might be required on laptops in certain quasi-static environments such as offices or public buildings.

Current fixed wireless systems under development claim to be able to provide data rates of around 10–20 Mbps to the home. Wireless LAN systems also claim to be in a similar range. Hence, the 10-Mbyte file transfer rate may be met within the next five years or so if these devices are deployed in a widespread fashion. Whether the 100-Mbyte transfer rate requirements will arise and will be met is less clear. Although there is little information available on how file sizes have grown historically, Figure 3.5 shows the growth in the average size of the hard disk, which is a reasonable proxy for the file size. In fact, there have been many different sorts of hard disks, for devices spanning mainframes to handheld devices. Here, to maintain a comparable basis we have plotted the average disk size in a PC against time.

What is clear from this chart is that when disk size is plotted on a logarithmic scale against time, there is an almost straight-line relationship, although the growth did slow a little around 1990. We might assume that this trend will continue onward throughout the period of interest, and using the trend line from 1990 to 2000 as the basis we can extrapolate to gain the graph shown in Figure 3.6.

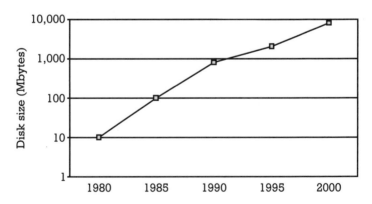

Figure 3.5 Average hard disk size on a PC.

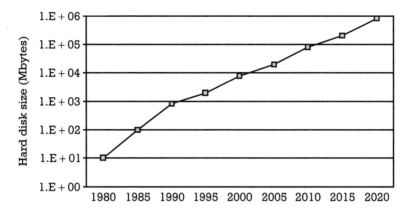

Figure 3.6 Prediction of future hard disk sizes.

On this basis we can expect hard disk size to rise by around an order of magnitude every 10 years, getting to around 800 Gbytes by 2020. Hard disk sizes only increase because users have a need for the capacity, although there is something of a circular issue here since as hard disk sizes increase, software producers write programs able to take advantage of the increased size; these larger files in turn spur additional growth of hard disk space. It seems only a matter of time before video editing capabilities are standard on PCs, which will result in huge video files containing an hour or more of video. At first, these will be archived to other storage media, but as time goes on, they will increasingly reside on computers. So, as a best guess, we might expect average file sizes on computers in 2020 to be 100 times larger than they are now. Users would like to be able to send these files through networks, although if the network capacity is limited, they will limit their transmission to smaller files. We have noted that users today often try to send 1-Mbyte files through modems—in particular, presentation files often take up this much space. By 2020, we might expect average users to be sending 100-Mbyte files routinely, with some high-end users wanting to send more data than this. As demonstrated above, reducing download times to almost unnoticeable levels would require 160-Mbps data rates. Users may tolerate longer times than this—perhaps up to a minute, if there is a key cost tradeoff. At a minute download time, the data rate requirement is 8 Mbps. Hence, we might predict that users in 2020 will demand an absolute minimum of 8 Mbps but would prefer rates of over 100 Mbps and may be prepared to pay more for this.

Potential issues for the future, then, are the following:

- The extent to which broadband fixed wireless networks will be deployed;

- The extent to which W-LANs will be deployed within private and public buildings and whether they will be widely accessible;

- The file size over the next 20 years and the speed with which users will want to transfer these files;

- Whether technology will be developed to cope with predicted growth in file sizes.

We recall part of Chapter 2's broad definition of what a user in 20 years might want:

Communication with any machine, potentially including machines not yet invented at the time the communicator was manufactured, in any meaningful manner for the human involving video, speech, and text.

In principle, we could communicate with a machine by connecting it to a network and defining a protocol to transfer information. Currently, we do communicate remotely with our telephone answering machine and are able to retrieve messages, change the greeting, and modify the mode of operation of the device. Of course, the answering machine has the rare advantage of already being connected to a communications network. Problems that need to be solved here are described as follows.

- *How to physically communicate with machines.* This will most probably be in a wireless fashion for maximum flexibility and may be through the BlueTooth technology.

- *How humans can communicate with these machines.* At present, the devices tend to use speech synthesis to "talk" to the human but accept input through a keypad. In the future, more advanced forms of interaction can be expected, including speech recognition and artificial intelligence.

- *What higher level protocols machines will use to talk to other machines.* On the face of it this appears to be a problem of the highest complexity.

Probably, for machine communications, bandwidth will be less important. It is hard to foresee the need for high-quality video when interacting with a machine, and for machine-to-machine transfer, very sparse protocols containing only raw information can be used. Hence, it seems unlikely that bandwidth will be an issue.

Again, consider part of Chapter 2's broad definition:

We could conceive of the ultimate communications network as one that:

▸ *Supports any number of the above devices;*

▸ *Allows any machine to communicate with any other machine in any meaningful manner.*

The first of these points raises immediate questions of network capacity. If the number of these devices rises in an exponential manner, as has the number of nodes connected to the Internet, then it is possible that network capacity may be rapidly consumed. Alternatively, if many of these devices transmit relatively little information, then the impact may be limited. Given the likely development of picocellular networks, on balance, capacity will probably not be problematic. The second point has already been dealt with when discussing machine-to-machine communications.

We will look at the key issues associated with each of the above points in Part IV, after having considered the constraints that prevent unlimited development and deployment of solutions to these problems. All the problems raised above need to be discussed, but by way of summary, the key issues are the following.

▸ Integration with the office and home wireless networks;

▸ How the mobile device will be linked into the computer system for the forwarding of e-mails given the rather different network concepts of voice and data systems;

▸ The economics of the provision of video communications to wireless devices;

▸ The extent to which broadband fixed wireless networks will be deployed;

▸ The extent to which W-LANs will be deployed within private and public buildings and whether they will be widely accessible;

- The file size over the next 20 years and the speed with which users will want to transfer these files;

- What higher level protocols machines will use to talk to other machines.

We will be examining these issues in detail in Part IV of this book. First, though, we turn to look at the constraints that may prevent change from happening.

References

[1]   Mouly, M., and M.-B. Pautet, *The GSM System for Mobile Communications*, published by the authors, 1992. See http://perso.wanadoo.fr/cell.sys/.

[2]   Harte, L., S. Prokup, and R. Levine, *Cellular and PCS: The Big Picture*, New York: McGraw-Hill, 1997.

[3]   Webb, W., *The Complete Wireless Communications Professional*, Norwood, MA: Artech House, 1999.

[4]   Webb, W., *Introduction to Wireless Local Loop*, Second Edition, Norwood, MA: Artech House, 2000.

[5]   Bing, B., *High Speed Wireless ATMs and LANs*, Norwood, MA: Artech House, 2000.

[6]   Held, G., *Data over Wireless Networks: BlueTooth, WAP, and Wireless LANs*, New York: McGraw Hill, 2000.

[7]   Clark, M., *Networks and Telecommunications*, Second Edition, New York: John Wiley, 1997.

[8]   Hanzo, L., W. Webb, and T. Keller, *Single and Multi-Carrier Quadrature Amplitude Modulation*, New York: John Wiley, 2000.

[9]   Moridera, A., K. Murano, and Y. Mochida, "The Network Paradigm of the 21st Century and Its Key Technologies," *IEEE Communications Magazine*, November 2000, pp. 94–98. (*See also* other related papers in this journal.)

[10] Zvoner, Z., P. Jung, and K. Kammerlander, *GSM Evolution Towards 3rd Generation Systems*, Boston, MA: Kluwer Academic Press, 1999.

What limits the future

CHAPTER

4

Overview of technology development

"Speaking movies are impossible. When a century has passed, all thoughts of our so-called 'talking pictures' will have been abandoned. It will never be possible to synchronize the voice with the picture."

—D. W. Griffith, film director (1874–1948).
Remark made in 1926.

4.1 Overriding principles of wireless communications

Wireless communications systems have reached incredible levels of complexity. The processing power in a mobile handset equals that in low-end desktop computers. The standards for a system such as GSM extend to many thousands of pages, and it is highly unlikely that there is anyone who understands all aspects of the GSM system in detail. Some engineers specialize in speech coders, others in radio frequency (RF) technology, and so on. Despite this complexity, there are a number of overriding principles and laws that can be used to understand the limitations of wireless communications. These are critical because they will shape the capabilities and economics of future wireless systems. This section distills the key issues and discusses their implications for future wireless systems.

As an overview, some of the guiding principles that determine the shape and form of all wireless communications systems are described as follows:

- *Bandwidth is scarce.* Wireless communications requires radio spectrum. There is an almost unlimited number of potential wireless applications but a limited supply of spectrum. This is a critical issue that will be examined in more detail in Section 5.3, but, broadly, because bandwidth is scarce, it becomes expensive. Hence, the more bandwidth required, the higher the cost of transmission. The cost of providing bandwidth is falling, but the cost of electronic devices is falling even faster. Hence, wireless communications devices will utilize extreme complexity to compress and reduce the amount of bandwidth required. It is very important to note that this is in contrast to wireline communications where, in areas such as fiber optic communications, bandwidth capabilities are growing faster than requirements and hence the cost of transmission is falling rapidly. This is leading to ever-increasing differentials in costs of transmission per bit between wired and wireless networks.

- *Providing ubiquitous coverage is problematic.* Radio signals are attenuated or reflected by most obstacles. Providing good coverage requires many base stations. Even worse, coverage becomes increasingly problematic at higher frequencies as diffraction and reflection losses increase; however, the demand for increased bandwidth is moving wireless communications systems to higher frequencies where there is more available spectrum. This tradeoff between the cost of coverage and the cost of bandwidth by moving up the radio spectrum is one of the classic design problems for wireless engineers.

- *The radio channel is hostile.* Not only is it difficult to provide coverage, but even when achieved the quality of the radio signal will be poor compared to that received on wireline. As a result, complex systems such as error correction mechanisms, interleavers, equalizers, and adaptive antennas are required in order to ameliorate these problems.

▸ *Battery power is an issue.* Most wireless communication is directed to portable devices. Most users want these to be as small and light as possible but at the same time allow long operation times between recharging. Hence, the power consumption of the device becomes a key issue, and unfortunately battery technology has not evolved at anything like the pace of semiconductor technology. This tends to encourage designers to utilize lower power transmission and less powerful computational devices. This sets up another tradeoff between the cost of transmission and the utility of the device.

▸ *Devices are generally portable and need to be inexpensive.* This results in a need for highly integrated solutions made in very large quantities. This is one of the attributes of a successful standard, and so this requirement, among others, tends to force the development of worldwide standards.

4.2 "Laws" shaping technical development

There are a number of laws, theorems, and general guiding principles that shape what can be achieved with wireless communication systems.

4.2.1 Moore's Law and its impact on system improvements

Moore's Law is not really a law, rather an observation that has held true for some 20 years. Moore stated that the processing power of a microprocessor will double every 18 months. When plotted against time, this results in an exponential increase in processing power. The implications of this law on everyday life have been profound, with computers becoming rapidly more powerful and less expensive and processing devices being embedded in many appliances—even simple and inexpensive ones like toasters. For the designers of mobile radio systems, Moore's Law has enabled designers to broadly ignore the complexity implications of their design. For example, when the GSM system was first conceived in the mid-1980s, the designers were well aware that with the current technology it would be impossible to implement handheld devices. They were sure, however, that technology would improve within approximately two to four years to a level where handheld devices would become possible. They were correct, and their foresight allowed a substantially better and more efficient system to be designed with important implications for its future proofing and utility.

Moore's Law, however, applies only to microprocessors and other associated baseband circuitry such as digital signal processors (DSPs). It certainly does not apply to the efficiency of radio communications where, for example, the number of voice calls per megahertz of spectrum has changed at a pace much slower than the increase in processing power. This is something that we will return to in Section 4.2.2, but suffice it to say here that computers are evolving much faster than wireless systems, which has worrisome implications in terms of the ability of wireless communication to enable computers to be connected into networks.

One of the implications of Moore's Law has been that for a given processing power, processing circuitry has rapidly fallen in cost. This has not been the case for radio frequency equipment, resulting in RF equipment taking an increasing share of the cost of the device, especially with the move to higher frequencies where RF equipment becomes increasingly expensive.

The net effect of Moore's Law is that, as far as the future is concerned, it can be assumed, reasonably safely, that no baseband processing will be too complex to be realized in the next 20 years. If Moore's Law holds over this time, then processing devices will be nearly 10,000 times more powerful (or cost commensurately less) than they are today. Implementation will not be limited by device constraints but by the capabilities of humans to develop algorithms of sufficient complexity that such processing power is required. Again, it cannot be overemphasized that an improvement in processing power does not automatically imply an improvement in the efficiency of radio transmission. Enhanced processing power can be used to implement, for example, better voice codecs, which will reduce the transmitted bit rate and hence increase the system efficiency. Typically, however, the gains are hard won, so a thousand-fold increase in processing power might halve the coding rate and hence double the efficiency. This improvement in efficiency is a very good thing, but the improvement is dramatically less than the increase in processing power used to realize it.

The same policy of "do not worry, it will be developed" cannot be assumed for RF systems. Here, costs fall slowly and it is certain that the RF systems of 20 years from now will not be 10,000 times "better" than those of today. Perhaps they will only be 10 times better, or maybe even less. These are important design considerations to bear in mind when considering the capabilities of devices of the future.

4.2.2 Shannon's Law and the enhancement of spectrum efficiency

It has been stated that one of the key principles of wireless communications is that bandwidth is scarce. There are widely agreed scientific principles in place that set out just how scarce bandwidth actually is; preeminent among them is Shannon's Law [1, 2]. This law states that under certain assumptions, the maximum information measured in bits that can be transmitted per second per hertz of radio spectrum is:

$$C = \log_2 (1 + \text{SNR}) \tag{4.1}$$

where SNR is the signal-to-noise ratio at the receiver. Practical systems approach to within around 3 dB of the Shannon limit for high-complexity receivers, which allows the limit to be used as a realistic upper bound on performance. In a tightly clustered situation, however, the performance of the system tends to be dominated by the signal-to-interference ratio (SIR) rather than the SNR. Gejji [3] and Webb [4, 5] performed some work in this area relating to cellular systems. Gejji started with Lee's equation [6], relating the number of radio channels to SIR, and given by:

$$m = \frac{B_t}{B_c \sqrt{\frac{2}{3} \text{SIR}}} \tag{4.2}$$

where B_t is the total available bandwidth, B_c is the bandwidth required per call, and m is the number of radio channels per cell. Gejji then replaced SNR in Shannon's equation with SIR and substituted for B_c within Lee's equation to derive:

$$m = 1.224\alpha \frac{B_t \log_2 (1 + \text{SNR})}{R_b \sqrt{\text{SIR}}} \tag{4.3}$$

where R_b is the user bit rate and α is a factor relating to the closeness to which the Shannon limit can be approached. Webb then showed that this could be optimized by using variable rate modulation at SIRs of around 6 dB. This leads to:

$$m = 1.42\alpha \frac{B_t}{R_b} \tag{4.4}$$

In the perfect case when the Shannon limit is reached, α becomes 1. This equation then tells us that if we have 1 MHz of spectrum and each user's "call" requires 10 Kbps of data, than we can accommodate approximately a maximum of 142 voice calls per cell in a clustered environment where spectrum is reused from cell to cell. Hence, unlike Moore's case, where things just keep getting better, with Shannon we are only able to approach some maximum limit. At present, we are still some distance from this limit. Proposed third-generation systems are expected to realize around 60–100 "calls" per carrier per sector. Each carrier is 5 MHz wide, so the number of "calls" per sector per megahertz is around 12–20. By adding more sectors it might be possible to reach levels of around 30–50 calls per cell.

The fact that we are some distance from the Shannon limit is not unreasonable. Shannon assumes perfect Gaussian channels, whereas mobile radio channels are considerably more hostile, with fading and multipath propagation [7]. Overcoming these problems requires error-correction coding, which typically reduces the efficiency to around half of its original levels. Other problems such as device imperfections, complexity of filtering, and inability to implement perfect receivers all add to the difficulty in reaching this limit.

It is not particularly relevant for envisioning the future that we determine exactly how closely we might approach to this limit. As we will see, the requirements for bandwidth are so great that the existence of this limit and the fact that we might, at best, only triple the efficiency that we have today, indicates that other mechanisms will be required in order to provide the bandwidth we will need. Actually, at this point, we have not yet defined how much bandwidth will be needed but have merely noted that for video transmission substantial increases in bandwidth over that provided by today's cellular systems will be necessary. We will return to calculate the bandwidth required in Part IV. Here we are interested in understanding the constraints that apply to providing such bandwidth.

4.2.3 Going beyond the Shannon limit?

Some recent work in Bell Labs and elsewhere has suggested that perhaps the Shannon limit does not apply to wireless transmission. A new approach called Bell Labs Adaptive Space Time (BLAST) was proposed by Foschini [8]. This approach is now generally referred to as multiple-input multiple-output (MIMO) antennas. The basic principle here is to have a number of antennas at the base station and a number at the subscriber unit. A different signal is transmitted from each antenna at, say, the base station, but all transmissions are at the same time and same frequency. Each antenna at the subscriber unit will receive a signal that is the combination of all the transmissions from the base station. Hence, if the signals transmitted through each of the base station antennas are S_1 to S_m and the radio propagation characteristics between antenna m at the base station and antenna n at the subscriber unit are C_{mn}, then the signals received at the subscriber antennas are given by:

$$
\begin{pmatrix} R_1 \\ R_2 \\ \dots \\ \dots \\ R_n \end{pmatrix} = \begin{pmatrix} C_{11} & C_{21} & \dots & \dots & C_{m1} \\ C_{12} & \dots & \dots & \dots & \dots \\ \dots & \dots & \dots & \dots & \dots \\ \dots & \dots & \dots & \dots & \dots \\ C_{1n} & \dots & \dots & \dots & C_{mn} \end{pmatrix} \bullet \begin{pmatrix} S_1 \\ S_2 \\ \dots \\ \dots \\ S_m \end{pmatrix}
\tag{4.5}
$$

where R_n is the signal received at the nth antenna. If all the parameters C_{mn} are known, then it is relatively simple to solve (4.5) by multiplying the received signals by the inverse of the channel matrix C_{mn}. In principle, by using this technique it is possible to achieve an infinite increase in capacity by using an infinite array of antennas. Each path from an antenna to another conforms to the Shannon limit, but the combination of all the paths can rise many times above it. The simple reason why this exceeds Shannon is that there are multiple paths involved here, not the single path that Shannon envisaged, although each path is not orthogonal to the others. This much the research community is agreed on. The key question is whether it will actually work.

In order to work, there must be a reasonable degree of difference between each of the channel characteristics C_{mn}—if they are all identical, then each of the signals received at the subscriber unit antennas will also be identical and it will not be possible to extract the different transmitted signals. This implies that a rich fading environment, as found in the world of mobile communications, is required. It is also necessary, however, that each of the parameters C_{mn} is known with high accuracy, otherwise the solution to the matrix equation will be inaccurate. In a typical mobile environment, the parameters C_{mn} are constantly changing as the mobile moves, or the scattering environment changes. Keeping track of the changes in these parameters without spending significant resources sounding the channel will be challenging. Indeed, the more diverse the parameters C_{mn}, the more rapidly the channel is likely to be changing. Based on experience in attempting to use channel measurements in repeaters and in the interference cancellation concept for CDMA, it seems likely that it will not be possible to track channel parameters with enough accuracy in a mobile environment for a large number of high data rate streams to be transmitted simultaneously. In the fixed environment, tracking the channel parameters is possible, but there may not be sufficient diversity across the channel parameters for the concept to work.

Hence, here we draw the tentative conclusions that MIMO will not result in significant capacity gains over and above those predicted by Shannon. This still remains to be demonstrated and is one area for potential benefits in the capacity equation.

Given that we are pushing up against limits in terms of the number of "calls" per cell per megahertz and the scarcity of spectrum is increasing and so there is a limit to the number of meghertz assigned to an operator, then the single parameter left is the number of cells. This fact was realized many years ago by Steele [9], who postulated that the basic principles outlined above would inevitably lead us to the deployment of smaller and smaller cells, until we had reached the stage where we had a microcell mounted on every lamppost, covering around 100m of the street, and picocells located in every room. This is increasingly happening, and it is clear that his vision was particularly farsighted. We do now have microcells in major city centers, perhaps not yet on every lamppost, but certainly of a high density, with, for example, 500–1,000 per operator in the center of Paris and similar numbers in other capital cities. We do not have much in the way of indoor cells yet, however; as already discussed in Chapter 3, these might be implemented soon in many ways, including W-LAN and PAN implementations.

Technology tells us that small cells are going to be essential in the realization of the wireless communications future set out in Chapter 3. Small cells, however, bring with them many problems:

- The base station equipment needs to be small and inexpensive.

- Linking the base station equipment back to the network needs to be simple and cost-effective.

- The complexity of managing numerous base stations in areas such as frequency assignment needs to be overcome.

- Handover and other mobility features become more critical in small cells as less time is spent in each cell.

Because of the critical role of small cells in providing enough capacity to drive our vision of the future, it is worth spending some time looking at each of these factors. This is the subject of Section 4.3.

4.3 The problems associated with small cells

4.3.1 Size of base station equipment

Until recently, cellular base station equipment has been large. A typical base station took up the better part of a room. Each base station had a number of sectors with multiple channels on each sector. Each channel had the capability to provide multiple voice calls. The transmit power was high, necessitating a separate rack of RF equipment. Large power supplies were required. This was a serious problem, especially when considering reducing the size of the cells. All this equipment needed a lot of space, and space was expensive to rent. As can be imagined, the equipment itself was also expensive.

This paradigm started to change around the late 1990s when cellular manufacturers started to make microcell base stations. These were typically the size of a suitcase and contained within them the power supply and other ancillary equipment. To achieve this reduction in size, much simpler devices were used, offering only one or two carriers and no sectorization. Maximum transmit power levels were much lower, perhaps only 100 mW to 1W, simplifying the RF equipment and the power supply. These units were typically deployed on the internal basement wall of buildings, with cabling being run to an external antenna. Some units were suitable for external mounting, but this was often problematic, as it was difficult to find a site where such a device could be mounted externally without being somewhat noticeable.

The next change was the design of picocellular systems. These were indoor mounting units with only a single carrier and power levels in the region 10–100 mW. By 1999 these units were around the size of a large textbook and were mounted on internal walls. They had integrated antennas and needed connection only to the standard building power supply and a communications channel such as a standard twisted pair socket.

With the introduction of third-generation mobile radio systems, this trend may be somewhat reversed, at least at first, since third-generation systems are more complex, with wider bandwidth carriers that have inherently much more capacity than a second-generation carrier. In the longer run, Steele [9] has postulated the development and deployment of base stations no larger than a coffee cup, which could be easily assimilated into the building or urban fabric. There seems to be little reason why this goal cannot be achieved in the relatively near future. Simple base stations are little more than high-performance subscriber devices, and so it should be possible to engineer them to a similar size. Perhaps the key paradigm that needs to be overturned to achieve this is the thought that base stations are different from mobiles in the volumes and difficulty of production and so are developed on a different kind of product line, less geared to mass production. Once base stations are seen as mass-market devices to be built in the millions, then the design paradigms and the price will change.

However small they become, base stations are highly unlikely to be as inexpensive as mobiles. This is because mobiles are somewhat simpler than base stations and will always be made in larger volumes. The end-user price of low-end mobiles is rapidly falling to $100 and will probably go to perhaps $50 over the next 20 years. We can probably bound the lowest cost of base stations as falling to the order of $100 each for simple picocells in the next 20 years. Of course, this will not be an accurate prediction, but it is the order of magnitude that is of interest here, rather than precise numbers. It seems highly unlikely that base stations would reach $10 as they contain too many parts for that. Equally, it seems almost certain that they will fall below $1,000—they are already rapidly approaching this point. So, from an order of magnitude prediction this is fairly certain.

Cellular systems are not the only type of base station equipment that might be of use. Within the home and office we have already noted that W-LANs and PANs might provide wireless communications. At a first glance, W-LAN base stations are not dissimilar from picocellular base stations in their capacity, output power, and size. At present, the lack of voice capability tends to simplify them slightly, but as voice becomes a part of W-LAN traffic this distinction will disappear (either the W-LAN systems will get voice cards or the cellular systems will become packet-based carrying VoIP). Hence, at the overview level, we might assume that W-LAN and cellular systems have similar base station size and cost.

Initially it might appear that PANs are different. As discussed in Section 3.2.5, BlueTooth chips are designed to cost around $5 each by 2005. A BlueTooth chip alone, however, does not make a base station. In addition, the chip requires housing, power supplies, and a means to connect the ad hoc network back to the building network, such as the HomePNA card discussed in Section 4.3.2.3. It needs additional protocols in order to talk with the network. It may need to be slightly higher power in order to have a greater range. Its one key advantage is that, if BlueTooth is successful, there will be enormous economies of scale. As a result, we might expect that a node of this sort might be around half the price of a cellular base station, perhaps around $50 per node. For a typical house with perhaps six rooms that require coverage (assuming that the garage, bathroom, and so forth are not critical and probably get covered from nearby rooms), the cost of building this sort of wireless communications system in a home might be around $300. Although not trivial, this is in line with the cost of other consumer devices such as televisions and video recorders.

In summary, it looks as though problems related to the size and cost of the base stations might be solved by 2005, with base stations being small enough to be easily hidden and costing around $50–100.

4.3.2 Backhaul

Base stations need to be linked back into networks. Only in very rare cases are they of any value if they are not connected to anything. Backhaul is one of the key problems for today's cellular operators, resulting in substantial cost and effort. A typical base station today requires a backhaul capacity of around 2 Mbps to the base station controller, which may be 50 km or so away. There are a number of ways that this backhaul could be achieved:

▸ *Installing a copper or fiber optic cable.* This is rarely done because the cost of installing cabling is becoming prohibitive [10].

▸ *Leasing capacity on a cable from a company providing such a service.* Often the company providing the service is the PTO. This is less expensive, although if the base station is in a remote location (such as atop a hill), there is often a large up-front payment to cover the cost of running a spur from the existing cable to the new site.

▸ *Utilizing wireless backhaul,* typically using point-to-point or point-to-multipoint fixed wireless equipment of the kind discussed in Section 3.2. This is often preferred as it is both relatively inexpensive and does not place a reliance on another telecommunications operator who is probably a competitor.

▸ *Utilizing existing wiring.* As DSL standards and cable modem standards have improved the performance of twisted pair and cable lines, respectively, the possibilities of using these for backhaul have become greater, especially for base stations located in buildings.

The ideal backhaul system would be free and available everywhere. That is never going to be the case, but clearly backhaul has the potential to destroy the business case for the deployment of multiple small cells. Indeed, today, it is often the largest single obstacle in the deployment of microcells. For the purposes of the discussion here, we do not need to focus on backhaul of cells in remote locations. These cells are generally already installed with existing backhaul. Instead, we are concerned with the provision of backhaul to small cells in areas where there will be high traffic levels, such as urban areas, buildings, and roadways. It is worth looking at a number of different examples here to further understand the problem, namely:

▸ A microcell on a lamppost;

▸ A microcell within a commercial building;

▸ A picocell within a home.

That is not to say that these are the key, or only deployment scenarios, but that by examining these we will gain a good understanding of the problem, which we can apply to apply to other scenarios.

4.3.2.1 Microcell on a lamppost

Each lamppost will always have a power supply. It is possible but unlikely that it has a communications connection already. It is highly likely that there will be buried copper cable in the vicinity of the lamppost; indeed, the cable may share the same duct as the power supply to the lamppost. Another advantage of a lamppost is that it is relatively tall and so rises above much of the urban clutter. It is therefore possible that a LOS radio link might be established with a radio backhaul network, perhaps operating in the frequency bands above 20 GHz. Realistically, the options open to backhaul this microcell are the following:

▸ *Utilize the power supply cable.* To date, efforts to send high data rate signals along power supply cabling have met with many problems.

It is possible that this may improve in the future, but given that many base stations might be connected to the same cable, and the bandwidth that they require will probably grow rapidly, then this seems unlikely to provide a viable solution.

▸ *Utilize the twisted pair wiring.* If an unused twisted pair can be found, then a base station can utilize techniques such as high-speed digital subscriber line (HDSL) to provide a 2-Mbps symmetric link along the cable. The operator will probably have some digging and climbing to do in order to connect the cable with the top of the lamppost and will have to pay the owner of the cable a rental for its use, but otherwise this is perfectly feasible.

▸ *Run their own wiring.* This is generally problematic. If there is room in the ducts running under the lampposts, then it might be possible to do this relatively cost-effectively, but if substantial digging effort is required, then the costs will rapidly get out of control.

▸ *Utilize a radio backhaul.* This would need a system with capacity levels of the order of 20–200 Mbps to handle a number of base stations in the area. These systems typically only exist in the frequency bands above 20 GHz. The problems here are that equipment is expensive (much more expensive than the projected cost of the base station) and reliability can be poor in conditions of high precipitation.

Although some manufacturers have proposed the use of 40-GHz links from the microcells, in practice, 40-GHz links cost around $5,000–10,000 per link and are unlikely to fall dramatically in the future since the main cost component is the RF equipment, which, as we have already discussed, does not follow the same pricing trends as the baseband equipment. This seems somewhat expensive if these base stations are to be deployed on many lampposts. All that remains is the use of the twisted pair. To utilize leased twisted pair, HDSL modems will be required at each end of the link, costing perhaps $1,000 but falling rapidly as HDSL modems are composed only of lower frequency components. A rental of perhaps $20–40 a month might then be charged by the company owning the twisted pair, with the amount depending on the country and the pricing trends on leased lines (which are invariably downward). This is a viable solution but one that will not work in all areas, and it is a long way from the free backhaul that was the requirement.

4.3.2.2   Microcell within a commercial building

In this case, the microcell will probably be mounted on an internal wall. Most commercial buildings have comprehensive in-building wiring and this would form the obvious backhaul for the microcell. The wiring might be twisted pair, Ethernet LAN, or fiber optic or coaxial cable. All would provide the capacity required and would be suitable. These solutions, however, only backhaul the microcell to a central point within the building. The signal still needs to get from here to the operator's network. This central point will almost certainly already have a communications link back into a network (probably the PSTN), which is how the current telephone traffic within the building is connected to the network. This link may or may not go to the desired location, and it may or may not have sufficient spare capacity to carry the base station traffic. Typically, neither applies, so it becomes necessary to lease another line. At present this is an E1 or T1 connection, which are expensive entities, costing as much as $1,000 per month in most countries. It is now feasible, however, to utilize a twisted pair with HDSL modem signaling at a much lower rate, if the leased line provider is prepared to provide it. This sort of backhaul will probably cost around $100 per month for the entire building. That is, all the microcells in the same building utilize this single connection. On a per-cell basis, this is much less expensive than the lamppost case.

4.3.2.3   Picocell within a home

Cost within the home environment is even more important than cost within the commercial environment. There are a number of mechanisms proposed for distributing signals around a home:

> *The IEEE 802.11 standard for W-LAN.* This is intended for the office rather than the home, but it could be used residentially as a means to backhaul picocell nodes. It could also be used as the primary radio link for the whole house, removing the need for a piconode in each room. The latter is unlikely, however, as it is somewhat expensive to be embedded in all home devices. As a backhaul it is plausible but perhaps a bit costly at around perhaps $200 per node. In a typical house with six rooms, this is around $1,200.

> *The HomeRF standard.* This standard is supposed to be to homes what IEEE 802.11 is to businesses—that is, a home communications system that links computers and other equipment together. It is designed to be a lower cost than IEEE 802.11 but is still unlikely to be

embedded in all devices. It is probably a better selection for backhaul than IEEE 802.11 since nodes are expected to be somewhat cheaper, perhaps only $100 per node.

› *The HomePNA standard.* This is a standard for the transmission of high data rate signals through the twisted pair wiring in the home. This is another strong possibility since the home picocells could be connected to a nearby telephone socket in each room and could run over the telephone wiring. At around $50 per node, prices for this are likely to be even lower than the HomeRF pricing.

› *The power wiring.* As mentioned before, it is possible in principle to send information over the ac wiring in the home. This would have the advantage that the piconode would need no additional connection beyond being plugged into a power supply outlet. As mentioned earlier, however, there are numerous problems with powerline transmission, and it is difficult to see all these being overcome—especially considering that people have been working on them for the past 10–20 years.

In summary, then, the HomePNA looks to be the most likely candidate for backhauling picocells within the home. Radio solutions could be used where there is difficulty in utilizing the home wiring, but HomePNA would be preferred.

As can be seen from the above discussion, there is no panacea to backhaul and no real likelihood that one will be invented in the next 20 years (because there are no other media over which the signals could be sent and key breakthroughs in the areas listed are unlikely). It is not a fatal flaw in the use of small cells, rather a cost that needs to be factored into the economics of deploying small cells. For cells within buildings, the problems are less severe. For microcells outside of buildings, backhaul is more problematic and is always likely to impose a significant cost burden on the microcell.

4.3.3 Management complexity

Managing a cellular system today is a complex business. Cellular operators typically have a number of operation and maintenance (O&M) centers where they can monitor every base station in the network. Here they can both check that it is operating correctly and modify the network by changing the frequencies assigned to the base station, thus changing its identity. As the number of cells in the network grows to well over 1,000 in many countrywide networks to date, the management of these networks becomes highly complex. In particular, each cell needs to be assigned a frequency such that the reuse of the frequencies is maximized but there is not excessive interference between any two cells. Cell boundaries need to be set such that there is sufficient overlap for handover to occur but not so much that frequency reuse is compromised.

The complexity of this operation increases, probably exponentially, with the density of cells in a particular area. As the cell count increases, the probability for interference increases, the number of neighboring cells increases, and the frequency reuse tends to fall. Some cells may be in tall buildings, necessitating three-dimensional modeling. Most operators today cannot conceive that they could successfully manage more complex networks, and for this reason are wary about the deployment of more cells, especially indoor cells, which are harder to model since often little is known about the building fabric and its propagation properties.

The solution to this problem is to utilize computing power to find the optimal assignment and to manage the network. Today, manufacturers are offering computing platforms that can look at an already-functioning network and make recommendations as to how it might be optimized. It is only a matter of time before these algorithms become well enough developed to perform all the management functions of the network. At the same time, the base stations themselves are getting more intelligent and are increasingly able to determine based on local measurements which frequency they should use. By coupling these two together, future networks should become self-managing, with all the existing complexity taken away from human operators. We can expect this to happen in the next five to 10 years.

4.3.4 Handover and mobility

Small cells also bring problems for handover. Handover algorithms typically work by noting that the signal strength or signal quality in the existing cell is falling and looking for other nearby cells that could provide an acceptable quality. At the point that the signal from the existing cell falls below some threshold, they cause the target cell to become the new serving cell. Key in this process is a period where the signal from at least two base stations can be received and measured. With very small cells it is quite possible for one to be covering one room and another a corridor. The walls provide a good isolation between the two cells. When leaving a room, the signal from the first cell can fall very quickly, not allowing sufficient time to measure the second cell and threatening to drop the call. The only real answer to this problem is to speed up the algorithm as much as possible and to design the handover algorithm so that if it loses the first cell, it quickly looks around for a second cell and reestablishes contact. The network then needs to rapidly reconnect the call. Techniques and algorithms to achieve this are under investigation and can be assumed to be in place within the next five years.

4.3.5 Implications for the future provision of bandwidth

In this section, we have learned that if bandwidth requirements increase in the future, as we expect them to, then the only effective way to provide this bandwidth in most cases is through the deployment of small cells. Although there are currently many problems with small cells, most can be expected to be solved in the next five to 10 years. Backhaul, however, will always remain a problem and will add significant cost to a cell. We might predict cost curves along the lines of those listed in Table 4.1.

These costs are very difficult to predict and depend on many factors, but they will probably be sufficiently accurate to serve as a guide when we look in Part IV at whether all the different features that the user requires can be provided. There, we will bring together some estimates of bandwidth required, willingness to pay, and cost of provision to understand when it is likely to become possible to provide certain services.

4.4 Technical developments in wireless in the last 20 years

In Chapter 7 we will take a detailed look at the changes that have taken place within the last 20 years and whether these could have been predicted by an analyst in 1980. Here we just summarize the key technical

Table 4.1
Predicted Costs for Small Cells

Type of cell	Cost per cell in 2000	Cost per cell in 2010	Cost per cell in 2020
Lamppost microcell	$1,500 plus $50/month	$500 plus $30/month	$250 plus $20/month
Business picocell	$1,000 plus $40/month	$400 plus $15/month	$150 plus $10/month
Home picocell	$100	$50	$30

developments in order to provide an illustration of the speed of change in wireless in the last 20 years. Although there is no guarantee that the same speed of change will apply in the next 20 years, this is nevertheless a good guide to what might happen.

In 1980 radio communications were analog and cellular systems were in their infancy, with many requiring operator intervention to set up a call and for the calling party to have knowledge of the location of the called party. By 2000 cellular systems were digital, used massively more processing power than the systems of 20 years earlier, and were almost fully automated. Satellite phone systems had been recently introduced and early versions of phones able to access Internet content were available. However, as shown in Chapter 7, in terms of capacity and capabilities, the key parameters for the end user, the advances were relatively small. The capacity per cell increased by only a factor of three, a worthwhile gain, but nothing in comparison to the increase in size of, say, hard disk drives. The capabilities extended slightly beyond voice to low-speed data services, but for most these were too expensive and too cumbersome and voice remained the key application. Voice quality improved somewhat, although it was still noticeably inferior to wired services for many calls.

There were few technical advances that had not already been understood in theory. Some of the more complex technologies introduced in the mid-1990s such as CDMA had already been well understood 20 years earlier, although more powerful processors and some novel ways to deploy the technology were required before they could become commercial. Technologies such as OFDM, which were just being used by 2000, had actually first been suggested in 1962. Most of the groundwork for voice coding and error correction coding was similarly already in place. There was much useful work conducted but this was typically in finding ways to enhance previously understood techniques or implement them at lower complexity.

We might conclude that compared to the computer industry, the changes in wireless technology were relatively slow (although, compared to well-established industries like automobiles, the changes were very rapid). The key advances were in producing technologies and products that could meet mass-market price points, dramatically driving the penetration upwards. Sections 4.1 and 4.2 focused on the scale of changes that might be expected in the next 20 years and has predicted a similar level of magnitude of changes as took place in the last 20 years. While the past is not always a good predictor of the future, assuming that trends observed in the past will continue into the future in the absence of other evidence is probably a good starting point.

4.5 Summary of the key technical issues

The key technical constraints that have been discussed in this chapter are as follows:

- The key technical principles that dominate wireless communications are that bandwidth is scarce, providing ubiquitous coverage is problematic, the radio channel is hostile, battery power is scarce, and devices generally need to be portable. These drive many of the solutions and constraints in mobile communications.

- There are a number of laws and observed principles that shape the way that technology evolves and is able to evolve. Moore's Law tells us that processing devices will rapidly become more powerful and less expensive. Shannon's Law tells us that there is a limit to the amount of information that can be transmitted through a mobile radio channel and that there is a maximum gain of perhaps three times that can be achieved over the capacity estimated for third-generation systems.

- Because of the limited capacity per cell, small cells will need to be deployed in order to provide the overall system capacity required. These bring a range of problems. Small base stations are required, and it is estimated that these might become available by 2005 at under $100 each. Backhaul of the information from the base station is needed, which will be an ongoing problem throughout the next 20 years, especially for small cells outside of buildings, and this will limit the deployment of small cells somewhat. The complexity of managing a very large number of cells will be steadily reduced within the next five to 10 years with automated network management packages, as will solutions to managing mobility within a small cell environment. As the result of all these issues, we expect small cells to become increasingly common over the next five to 10 years, especially within buildings.

‣ A study of the last 20 years suggests that a capacity increase of three times per cell over the next 20 years would be in line with the last 20 and that compared to computers, the pace of evolution of mobile communication systems can be expected to be relatively slow.

References

[1] Shannon, C. E., "A Mathematic Theory of Communications," *Bell Systems Technical Journal*, Vol. 27, July 1948, pp. 379–423, and October 1948, pp. 623–656.

[2] Shannon, C. E., "Communications in the Presence of Noise," *Proc. IRE*, Vol. 37, January 1949, pp. 10–21.

[3] Gejji, R. R., "Channel Efficiency in Digital Cellular Communications Systems," *Proc. 42nd IEEE Vehicular Tech. Conf.*, Denver, CO, 1992.

[4] Webb, W. T., "Modulation Methods for PCNs," *IEEE Communication Magazine*, Vol. 30, No. 12, December 1992, pp. 90–95.

[5] Webb, W. T., "Spectrum Efficiency of Multilevel Modulation Schemes in Mobile Radio Communications," *IEEE Trans. on Comms.*, Vol. 43, No. 8, August 1995, pp. 2344–2349.

[6] Lee, W. C. Y., "Spectrum Efficiency in Cellular," *IEEE Trans. on Vehicular Tech.*, Vol. 38, No. 2, May 1989, pp. 69–75.

[7] Webb, W., *The Complete Wireless Communications Professional*, Norwood, MA: Artech House, 1999.

[8] Foschini, G., et al., "Layered Space-Time Architecture for Wireless Communications in a Fading Environment When Using Multiple Antennas," *Bell Labs Technical Journal*, Vol. 1, No. 2, Autumn 1996, pp. 41–59.

[9] Steele, R., "Mobile Communications in the 21st Century," Chapter 11 in *Communications After AD2000*, Davies, D., C. Hilsum, and A. Rudge (eds.), London: Chapman & Hall, 1993.

[10] Webb, W., *Introduction to Wireless Local Loop*, Second Edition, Norwood, MA: Artech House, 2000.

Contents

Nontechnical constraints on the future

"There is nothing more difficult to take in hand, nor perilous to conduct, or more uncertain in its success, than the introduction of a new order of things, because the innovator has for his enemies all those who have done well under the old order of things, and lukewarm defenders in those who may do well under the new."

—*Niccolo Machiavelli (1469–1527)*

5.1 Introduction

In addition to the technical laws and problems discussed in Chapter 4, there are many nontechnical constraints and bodies that limit the speed of change that can be achieved. These include the areas of standardization, radio spectrum allocation, the financial needs of investors, and legal issues. All are discussed in this chapter. Of course, one of the key constraints is the actions of those who invent the future, and this is discussed in Chapter 6, which considers the ability of organizations to develop the future.

It is important to understand all these constraints in order to filter the broad user requirements that were developed in Chapter 2 so as to be able to understand the likely timeline over the next 20 years. The constraints effectively provide information on the speed with which we can move into the future, and in some cases prevent particular visions from happening.

5.2   Standards [1]

5.2.1   Introduction

Standards are critical to many of the different parts of wireless communications. For any device to be able to talk to any network, both the device and the network must communicate using the same protocol. This protocol needs to extend through the open systems interface (OSI) stack from, at least, layers 1 to 3. Specifically, it needs to address the following:

> ▸ The physical transmission including the frequency band to be used, the data rate, the bandwidth, and the power control;

> ▸ The layer 2 aspects such as the error-correction mechanism to be adopted;

> ▸ The layer 3 protocol aspects such as the mobility management, connection control, and radio resource management issues.

Beyond this, additional protocols will be required for aspects involving machine communications so that the machines know how to interpret certain commands in a uniform and universal manner. Without these standards there is simply no way that an individual's wireless requirements can be met in full. Instead, the solution will be limited to certain areas of those requirements where the devices or machines are able to communicate because they happen to share a common manufacturer that has enabled interconnectivity.

Regardless of the interconnectivity issues, standards also enable increased competition, which lowers prices and speeds acceptance. Many nonstandardized solutions have failed simply because operators are not prepared to accept a solution that is only supplied by a single manufacturer.

The need for standards has long been recognized within the mobile wireless communications industry. Standards for devices such as pagers and private radio systems have been developed by bodies such as the Conférence Européenne des Administrations des Postes et des Télécommunications (CEPT) since the 1970s, with standardization coming to the fore with the production of the GSM specifications by the European Telecommunications Standards Institute (ETSI) from around 1984. Standards are required in the mobile industry so that devices can roam from one network to another and still work, and the GSM standard enabled mobile phone users to roam throughout initially Europe and then much of the world with a single mobile terminal. Within the fixed wireless industry, standards have been less important because, by definition, both ends of the link are fixed and hence there is no roaming.

5.2.2 Advantages and disadvantages of standards

Aside from the obvious need for standards for some parts of wireless communications, it is worth understanding their advantages and disadvantages. The advantages of standards are described as follows.

‣ Standards provide a means whereby devices from different manufacturers can communicate with each other or with a common network.

‣ The standards process typically provides a level of public scrutiny that tends to reduce the likelihood of errors in the standard and, even better, increase the diversity of those working on the standard, leading to more views of the future and thus, typically, more future-proofing being built into the device.

‣ By bringing together many manufacturers and operators, the standards process itself tends to produce greater commonality in a vision of the future among all these different players, resulting in their actions being more coordinated and hence there being fewer "battles" of vision. This results in the future being realized more rapidly and smoothly than might otherwise have been the case.

‣ Standards provide operators and the public with the capability to multisource equipment, which drives down prices, results in more aggressive equipment development, and gives the users increased confidence that they will not have to purchase from a particular supplier indefinitely.

The disadvantages of standards are described as follows.

▸ The standards process itself is time-consuming. This is because there is a need to gather together a number of individuals, achieve consensus, subject the output to public scrutiny, and follow formal and clear procedures.

▸ The standards process can sometimes result in overengineered equipment. Often the simplest way to resolve a dispute concerning, for example, the likely requirements of a user is to build into the system the capability of addressing both conflicting views. As this is repeated over and over again, the equipment acquires capabilities that are highly unlikely to be needed, which adds cost and complexity.

Standards have some characteristics that tend to increase the development time and others that reduce it. It is worth looking in more detail at the time scales associated with standards because this is clearly a key factor influencing the speed of change. There are a number of stages to developing a standard:

▸ Formation of the initial idea for the need for the standard;

▸ Gathering of a consensus of industry and academia and persuading an appropriate standards body to formally start the process;

▸ Defining the requirements for the device;

▸ Defining the architecture of the device;

▸ Defining the detailed device specifications;

▸ Publishing the specification and reviewing comments;

▸ Establishing a type-approvals body (if appropriate);

▸ Monitoring and modifying the standard as problems arise or as enhancements are required.

Different standards have different levels of complexity and take different amounts of time to complete. The GSM standards took around six years for the first phase of the standards, and eight years until all the work originally envisaged was finally completed with the publication of the Phase 2 GSM specifications in 1992. Since then, the GSM committees have continued to meet to enhance the GSM specification. BlueTooth standards will probably take two to three years to complete but are generally simpler than systems such as GSM and involve far fewer interfaces and hence fewer protocols to define.

It is hard to provide any detailed estimates of the time required for standardization since this will depend on the complexity of the standard, the number of parties involved and their degree of divergence, and the exact standardization process. In practice, it is rare that equipment becomes commercial within four years of the point at which a standard is first contemplated, and it is certainly possible for the time scales to extend to six or even eight years. This, then, is clearly a key constraint on the future: If a standard is needed for a particular vision and work on that standard has not yet started, then it is unlikely that the vision will be realized within five years.

There are broadly two key models for the development of standards. The one described so far in this section is often thought of as the European model, as championed by ETSI, based on open forums and consensual development. The alternative is the one often adopted in the United States. Here, a company, such as Qualcomm, invents a wireless communications system without any discussion with other parties. When it becomes complete, the company goes to a standards body and requests that their finished system be given status as a standard. In order to grant this, the standards body will require that the company make available detailed system specifications such that others can build the system to the same specification. Other parties will then be invited to take part in committees that consider the appropriateness of the standard and whether changes are required. The key advantages of this approach are described as follows.

▸ It is typically much faster since only a small design team produces the initial specifications, and hence they do not need to develop consensus or travel to large group meetings.

▸ Because the system typically only has one chief designer, it is generally not overengineered to achieve consensus in the way that other standard-based systems can be.

The disadvantages of this approach are described as follows.

▸ The company designing the standard typically expects royalty payments from users of the standard, which tends to increase the cost of equipment.

▸ Not all companies will agree with the design approach, and so some may not be inclined to join in the standardization process, making participation less than universal.[1]

▸ The standard is often not subject to the rigorous checking that the other standards approach tends to produce.

It is still far from clear at this point which type of standards approach will prevail for wireless communications, or indeed whether the two will continue to exist in parallel. All standards bodies are trying to develop a model for standardization that brings together the best of these two approaches, and it may be time before we see a combination of the two. This might reduce the time required to develop standards perhaps to three to four years, but it is difficult to see much additional reduction given the complexity of the problem.

5.2.3 Relevant standards and standards bodies in the world of wireless communications

To some extent, there is only limited benefit to be achieved in listing all the relevant standards bodies. This is because many are formed for transient purposes, to achieve particular purposes. Others are being formed on an almost monthly basis, and so it is certain that by the time this book is published, the list will be incomplete. Nevertheless, a list of the key standards bodies to date does provide some useful indication of the breadth of the standardization work and the problems that might be anticipated in getting all the different standards to work together.

A detailed description of some of the key standards bodies involved in mobile communications can be found in [1]. Here we will just provide a brief description.

---

1. This is one of the reasons why the United States has a multiplicity of cellular standards, whereas Europe has only one.

Cellular radio

- ETSI provides standardization of many different telecommunications systems. In the mobile area it developed the GSM standards and is developing the Universal Mobile Telecommunications System (UMTS), one of the key variants of third-generation cellular. Although originally predominantly European, since the global success of GSM it now has a more global appeal and membership.

- The Telecommunications Industry Association (TIA) has been one of the key bodies for developing mobile standards in the United States through the TR-45 and TR-46 committees. These have developed the IS-136 (TDMA) standard and the IS-95 (cdmaOne) standard.

- The Alliance for Telecommunication Industry Solutions (ATIS) also has played a role in developing a GSM-based standard for the United States, but it is not a key driver for other standards development.

- The Association of Radio Industries and Businesses (ARIB) has been the key mobile standards body in Japan, developing the PDC cellular standard used widely in Japan. It is now seeking to work with other bodies, particularly ETSI, to develop a single standard that can be used around the world.

- Third-generation standardization also spawned a number of ad hoc bodies with tasks such as the third-generation partnership program (3GPP) and bodies developing mobile Internet standards, bringing together all the different standards bodies and more.

Private mobile radio

- Many of the bodies listed above also have divisions responsible for private mobile radio. In ETSI there is a group developing the terrestrial trunked radio (TETRA) standard, while within TIA, TR-8 develops U.S. private mobile radio standards.

Fixed wireless

- Within Europe, the Broadband Radio Access Networks (BRAN) committee within ETSI is responsible for developing fixed wireless standards and has been working on the HiperAccess standard. In 2000 it had just started looking at a standard for below 10 GHz.

▶ In the United States, the Institute of Electrical and Electronic Engineers (IEEE) standards body developed the IEEE 802.16 standard for fixed wireless systems above 20 GHz, and in 2000 it had just started a subgroup looking at sub-20-GHz standards.

W-LANs

▶ The key W-LAN standard was also developed by a branch of the IEEE and is known as IEEE 802.11.

▶ ETSI has also developed W-LAN standards known as HiperLAN and HiperLAN2.

PANs

▶ The key PAN standardization in 2000 was within the BlueTooth standards committee.

▶ The IEEE was seeking to standardize BlueTooth as IEEE standard 802.15.

▶ Work on next-generation BlueTooth solutions was taking place within both the BlueTooth and IEEE 802.15.3 bodies.

There are also other standards bodies involved in fixed communications, but these are outside the scope of this book. It is clear that there are many different standards bodies located in Europe, the United States, and Japan. To some extent, Japanese standards bodies are now aligning themselves with European bodies, but there still remain differences between the United States and Europe, resulting in different standards being developed in these two centers and two standards competing around the world. Within Europe, almost all standardization is controlled by ETSI, although the fact that two standards are being developed within ETSI for cellular and fixed wireless is no guarantee of the fact that they will interwork or use common design principles. Outside of Europe there are many diverse bodies that sometimes compete to be the key provider of standards in different areas. This leads us to conclude that there may be standards developed in most of the key areas, but that there may be more than one standard and that different standards in both the same and different areas are unlikely to interoperate. Section 5.2.4 considers whether any other standards might be required.

5.2.4   What other standards might be required?

In general, the following standards are available or under development.

Person-to-person:

- Voice communications systems such as GSM and cdmaOne and the third-generation systems such as UMTS and cdma2000;

- The capability for video communications within third-generation systems using the higher data rate modes;

- Support for text and image transfer within the second- and third-generation systems using the data modes available.

Person-to-machine/machine-to-person:

- Data modes within second- and third-generation systems;

- Local area network standards that allow a person to communicate to a computer network.

Machine-to-machine:

- BlueTooth as a means for machines to communicate over short ranges.

Backbone and other:

- HomeRF and other similar systems as a means for people and machines to communicate within the home environment;

- Fixed wireless systems as a way to link the home or office environment into external networks.

On the face of it, this would seem a comprehensive list of standards covering all the different areas that might be of interest. We might expect, however, that standards might be needed to cover aspects such as:

- The type of information that machines will communicate. For example, if the refrigerator is to tell the PDA that more milk is required, then the refrigerator must know how to tell any PDA to add "milk" to the shopping list stored in the PDA. Although BlueTooth may be sufficient to handle the lower three layers of the communications stack, there will be considerable complexity in the upper levels. This problem does not arise with person-to-person communications since humans can effectively manage the upper levels of the communications protocol. This is a complex problem because of the variety of different machines that might want to communicate and the wide range of different information they might want to send.

• The means by which different networks will communicate critical information such as the whereabouts of a subscriber.

▸ The means by which different standards will interoperate; for example, to hand over a subscriber from one technology to another such as from a W-LAN to a cellular system. There are many issues here including, for example, how to support a cellular phone on a W-LAN connected to a wired Ethernet backbone and then into the packet network.

We will discuss these issues further in Chapter 16.

5.2.5 The constraints that standards will bring to wireless communications

From the above discussion the following are apparent:

▸ If no standard is developed in a particular area, this will provide a severe constraint on the future, as standards are critical to enable widespread wireless communications. As the need for a standard, however, becomes increasingly apparent, generally one or more bodies become prepared to undertake that standardization.

▸ Standards take roughly four to six years. Hence, this is a severe and important limit on the speed at which change can occur.

5.3 Spectrum

It was mentioned in Chapter 4 that increasing the amount of information that could be transmitted was problematic. One way to increase the throughput of information would be to increase the spectrum available to the operator. Chapter 4 indicated that, in general, this is not possible. This section examined in more detail why this is the case.

It has been said before that "radio spectrum is like real estate—they just don't make it any more." It is a good analogy because radio spectrum cannot be created. Historically, users have increasingly used higher and higher frequencies in order to access unused spectrum. Moving to higher frequency bands also tends to result in the increased availability of spectrum—at around 1 GHz, 10% of the spectrum is 100 MHz; at 10 GHz, 10% of the spectrum is 1 GHz.

5.3.1 Issues in using higher frequencies

Moving to higher radio frequencies involves a number of important penalties.

5.3.1.1 Radio propagation is more problematic

As the frequency increases, the range is reduced, the losses from diffraction and refraction increase, and the ability of the waves to pass through objects such as buildings diminishes. As a rule of thumb, doubling the frequency tends to approximately halve the range. This results in the cell area falling to 25%, or four times as many cells being required to cover the same area. The issue of propagation is critically important to cellular systems where providing good coverage is one of the key differentiators for an operator. For other applications, such as indoor solutions, coverage may be less of an issue as there may be a separate picocell in each room, which has more than adequate coverage at the frequency of operation.

Over the last 50 years, the frequencies that were thought to be the highest useful frequencies for mobile radio communications have continued to increase, from 100 MHz to 400 MHz to today's cellular frequencies at 800/900 MHz and the PCS frequencies at 1,800/1,900 MHz. The frequencies proposed for third-generation cellular are slightly higher at around 2,000 MHz, with extension bands proposed up to 2,700 MHz. The degree to which the propagation penalizes the operator depends on the utilization of the network. In a typical deployment, operators first deploy sufficient cells to cover the area of interest. As the number of subscribers increases, the capacity required increases until additional cells are needed to provide sufficient capacity, at which point the range of the original cells is reduced in order to allow the insertion of additional cells. Once this starts to happen, the penalty for utilizing higher frequencies falls. This has been clearly seen with the PCS operators who, in principle, were hampered with reduced propagation compared to the cellular operators, but in practice, since the cellular operators needed to reduce their range to allow an increased number of cells to be inserted, the penalty has not been severe.

Given this apparent capability to increasingly move up through the frequency bands, it is worth asking whether there are any fundamental limits and where the various wireless communication systems will be operating over the next 20 years.

Cellular systems.   At present, the highest frequencies proposed for cellular use are 2.7 GHz, the extension bands for the third-generation systems. The key issue for cellular communications is that as the frequency rises, the need for LOS propagation increases. Limited measurements made to date have shown that above around 3.5 GHz it is very difficult to operate cells much larger than around 100m radius without LOS. Cellular systems will always need larger cells to cover rural areas, but as discussed in Chapter 4, it is possible that the use of lamppost-mounted cells and indoor cells will result in many of the cells having a range less than 100m. It would be possible for wide-area coverage to be provided in the current cellular frequencies and urban and indoor systems in much higher frequency bands. Actually, this is less important than might be imagined. Once cell sizes are down to very small levels, there are relatively large amounts of spectrum available since much of the total spectrum allocated can be used in a very small area. The net conclusion of this analysis is that it will probably not be possible to utilize spectrum much above 3.5 GHz for wide-area coverage, but that much higher frequencies, perhaps up to 10 GHz, will be used in small cells. Moving up to these frequencies, it is possible for an operator to have an allocation of at least 100 MHz, perhaps even 200–300 MHz. Assuming a reuse factor of, say, three, then anything from 30–100 MHz might be available in the future in small cells. The total available in large cells, however, will not be significantly different from that available today, certainly less than an order of magnitude improvement.

Fixed wireless systems.   Fixed wireless systems already utilize a very wide range of frequency bands from 400 MHz through to optical bands. Some of the key bands today are the 3.5-GHz bands and the so-called LMDS bands at 26–38 GHz. Typically, it is found that at 3.5 GHz it is possible to have non-LOS propagation as long as the subscriber is not on the edge of the cell. At frequencies above 20 GHz, LOS propagation is essential. As a result, the higher frequencies are typically only used in downtown areas where there is a high density of users and where the base station can be mounted on a high site with a good change of LOS propagation to a fair percentage of the buildings. At 3.5 GHz, operators typically have allocations ranging from 20 MHz to a maximum of 100 MHz. At 28 GHz, some operators have as much as 1 GHz of radio spectrum. They key issue here then, perhaps, is not how high a frequency will fixed wireless move to—once LOS propagation is accepted, then almost any frequency up to and including light is a possibility—but whether these higher frequencies can be used for more than business applications in the future. If they cannot, then fixed wireless services to residential homes are probably limited to a maximum of around 10-MHz bandwidth, allowing for reuse. If LOS propagation is acceptable, then 1 GHz of spectrum becomes a reasonable possibility. This is a massive difference with important implications. As discussed in Chapter 3, a possibility for enabling LOS propagation to work effectively with residential users is mesh technology, and we will return to this issue in Part IV.

W-LAN systems. These currently work in the unlicensed frequency bands at 2.4 and 5.7 GHz. We will return to the issue of unlicensed bands later in this chapter. With W-LANs, as with small cellular systems, propagation is not particularly problematic since numerous small cells can be installed and the range might only be a few tens of meters. As with cellular, it might be possible to move up to 10 GHz or perhaps even higher. As a result, an allocation of 1 GHz might be possible. The biggest problem for W-LAN systems has been obtaining allocations. Typically, W-LAN systems have not been able to obtain worldwide allocations and have instead opted for unlicensed spectrum, which is simple to obtain. There are, however, only a few unlicensed bands in use, and the largest of these is around 300 MHz (this is at 5 GHz, but the band is split across two allocations, which makes using the full allocation difficult). How much spectrum is actually available in each cell is more problematic to calculate since it depends on the reuse of unlicensed spectrum by other devices, as discussed in more detail below. Today's W-LAN systems are highly inefficient, providing only around 1 MHz of spectrum per cell, but much better systems might be envisioned that could provide up to, perhaps, 50 MHz in today's allocations, and even 200 MHz if larger allocations could be achieved in the future.

PAN systems. The same comments apply to PANs as for W-LANs since they both utilize the same spectrum. In principle, PANs have even smaller cells and so the spectrum can be reused more often. However, the uncoordinated nature of the reuse of the frequency might reduce the efficiency. This is difficult to determine, and so the best estimate may be to reuse the same numbers as for W-LANs—namely, 50 MHz to up to 200 MHz in the future.

One point worth making is that these wireless communications devices are not the exclusive users of radio spectrum. There are many other uses such as military, aeronautical, satellite, and radar, and so it is not possible to allocate all the spectrum to the applications listed here. Hence, the maximum allocation size listed here may not be achieved. Also, some of the bands discussed have been considered for more than one of the categories listed here. Generally, it cannot be provided to both, and so some sharing may have to take place. In summary, the maximum bandwidths that can be provided might be as shown in Table 5.1; the issues of how spectrum is allocated do not need to be understood here in order to determine the constraints, so they have not been listed. More details can be found in [1].

5.3.1.2 Devices are more expensive

Another penalty that is paid with increasing frequency is that RF devices become increasingly expensive. Table 5.2 shows the prices in 2000 for RF amplifiers at different frequency bands for typical subscriber devices.

It is difficult to construct a table that is simple to read because different power levels are available in different frequency bands, and at the higher bands, output power levels above 0.1W become increasingly difficult to find. For simplicity, however, if we extrapolate somewhat, we obtain the graph shown in Figure 5.1.

In Figure 5.1 the data points are somewhat variable, which reflects the volume of RF amplifiers currently sold in the band—as the volume increases, processes tend to be introduced that reduce the price of the amplifier. Nevertheless, there is a clear trend that can be modeled as an exponential increase in amplifier cost with frequency as modeled by the smooth curve in the graph. This curve is such that moving from

Table 5.1

Maximum Bandwidths in the Future

	Typical Spectrum per Cell		
	2000	2010	2020
Cellular	3 MHz	30 MHz in small cells 6 MHz in large cells	100 MHz in small cells 6 MHz in large cells
Fixed wireless	10 MHz in non-LOS 300 MHz in LOS	20 MHz in non-LOS 600 MHz in LOS	35 MHz in non-LOS 1 GHz in LOS
W-LAN	1 MHz	50 MHz	200 MHz
PAN	1 MHz	50 MHz	200 MHz

Table 5.2

Prices for Power Amplifiers in Different Frequency Bands in 2000

Frequency Band	Output Power Level	Typical Cost per RF Amplifier
2.5 GHz	0.1W	$12
	0.25W	$20
	1W	$50
3.5 GHz	0.1W	$14
	0.25W	$26
	1W	$53
5 GHz	0.1W	$24
	0.25W	$53
	1W	N/A
20–30 GHz	0.25W	$38
30–40 GHz	0.25W	$90
50–60 GHz	0.04W	$30

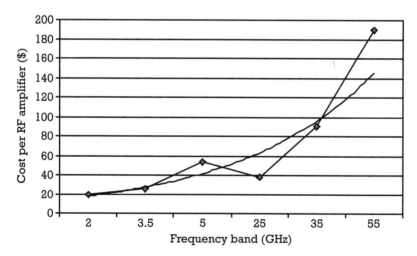

Figure 5.1   Prices for power amplifiers versus frequency band in 2000.

2 to 20 GHz results in a three times increase in RF amplifier costs. Other RF components such as mixers and low-noise amplifiers are also likely to see similar cost trends.

This problem can be reduced to some extent by using lower transmitter powers, but this further reduces the range, which has already been reduced through the use of a higher frequency. There is little that can be done about this other than to accept the penalty. Manufacturers then need to make the tradeoff of the utility of the increased bandwidth versus the higher cost. Generally, the option selected is either simply to utilize the greater bandwidth and hope that mass production will find a way to drive the prices down, or to develop devices at both frequencies and target them at different areas of the market (as has been done in the case of fixed wireless). Although the prices of the highest frequency devices have been falling, they have not been falling particularly fast, and they are unlikely to fall any faster in the future. This issue will keep most users below 10 GHz.

5.3.1.3 Antennas are smaller
Antenna dimensions are related to the wavelength of the transmission, which means that as the frequency increases and the wavelength reduces, the antenna dimensions tend to fall. This is normally desirable and is not typically a driving issue in the use of radio spectrum.

5.3.2 Unlicensed spectrum
Unlicensed spectrum is spectrum that can be used by anyone without a license. Since around 1995 unlicensed spectrum has played an increased role in radio communications. Unlicensed spectrum was initially developed for devices such as garage door openers and was often in bands where there were already industrial or other noncommunication uses. The 2.4-GHz frequency band is a prime example of this—the frequency 2.45 GHz is used by microwave ovens, and although they are well shielded, each one allows perhaps around 10 mW of radio energy to escape, which could act as serious interference to other communications. These unlicensed bands are often called "garbage bands" because of the variety of different systems that are thrown into the band.

Against this background it has been proving increasingly difficult to get radio spectrum allocations for new applications. Gaining an allocation can take many years and increasingly, as auctions are used for radio spectrum, can be very expensive. A good example of this problem is the use of fixed wireless frequencies in non-LOS bands in the United States. The 3.5-GHz band has not been allocated for this purpose, and so there was no obvious band. As discussed in [2], the MMDS band, a band originally envisioned for video distribution, became a possible band for this use when the Federal Communications Commission (FCC) allowed two-way communications in the band. Sprint and MCI WorldCom then spent around $2.5 billion between them buying many of the existing owners of the band in order to gain the spectrum for fixed wireless use. Other operators, such as Fuzion, based in Miami, have instead opted for the unlicensed spectrum at 5.7 GHz. Gaining access to this spectrum has not cost them any money, but they do not have exclusive rights to the spectrum.

W-LAN and PAN devices have tended to gravitate toward the unlicensed spectrum as a simple way to obtain allocations around the world. Most of the owners of these devices are not operators but are companies that want to install a W-LAN in their offices, or individuals who want to install PANs in their homes. Clearly, each of these is not going to apply for a spectrum license. An interim possibility would be for there to be an allocation for W-LANs that anyone could use, but only for W-LAN applications. This is the case for cordless phone systems, which tend to work in bands dedicated to cordless use alone. It is certainly possible that such allocations might be made for W-LANs and PANs in the future, although gaining a worldwide spectrum allocation typically takes around 10 years or more. Hence, we cannot expect to see this happen until toward the end of the time period under consideration here. For W-LANs, the building itself provides shielding from other users, and so the fact that the band is not dedicated to a particular user is not unduly problematic. For PANs, the PAN effectively follows the individual around, surrounding them in a "wireless bubble." This will clearly suffer interference from many other sources and will not have a natural protection around it. On the plus side, the range of a PAN is so small that PANs themselves will have limited interference from other PANs. Nevertheless, they can still be interfered with by a powerful source in the vicinity (e.g., the base station of a fixed wireless system).

For the next 10 years at least, then, we can expect W-LANs and PANs to have to use predominantly unlicensed bands, and for this to be a band also widely used by fixed wireless systems. The key issues surrounding unlicensed spectrum are the following.

- Zero cost of spectrum acquisition;

- Expectation of interference from other sources, which is nearly impossible to predict in the medium to long term;

- Potentially inexpensive RF devices because of the volume of other radio equipment being manufactured for the band.

Key here is the interference issue. With more and more devices targeting the unlicensed bands, it is possible that at some point the band will become unusable, dealing a major blow to the possibility of communicating in the future. Applications that are currently considering unlicensed bands include:

▸ W-LANs such as IEEE 802.11;

▸ Fixed wireless devices such as the adaptive broadband AB-access technology;

▸ PANs such as BlueTooth;

▸ Remote control applications.

This is in addition to existing interference sources such as microwave ovens. Furthermore, it must be remembered that there may be more than one instance of each of these different uses in the same area; for example, there may be competing fixed wireless operators using the same spectrum in the same area. Because of the critical nature of the possibility of all these interfering with each other, it is worth looking at the interference issue in more detail.

5.3.2.1 Interference to W-LANs

As mentioned above, W-LANs tend to have natural isolation in that they are almost invariably placed within buildings. Because the building environment is, to some extent, under the control of the building owner, the owner can, in principle, regulate the radio transmissions within that building. Because building penetration is relatively poor at the frequencies used for unlicensed transmission, unless there is a fixed wireless base station across the street pointing straight in the window, the possibility of serious interference originating from outside the building is relatively low. It is unlikely that there would be other W-LANs within the same building, or at least on the same floor. PANs, however, might provide a significant source of interference as there could be a PAN transmitter within a few meters of a W-LAN receiver. PANs are also difficult to control as they move with people. This is potentially a serious source of interference.

5.3.2.2 Interference to fixed wireless devices

Fixed wireless devices tend to be on the outside of buildings and so can be reasonably well shielded from devices within buildings. This may not be sufficient, however. A fixed wireless receiver pointing at a base station may point just over the top of a neighboring building. Within that building might be a W-LAN transmitter pointing straight out of the window and into the boresight of the fixed wireless receiver. Although the power level might be only 1% or 0.1% of that used by the fixed wireless transmitter, the proximity to the W-LAN might result in the interfering signal being stronger. If W-LANs are widely installed, then this starts to become a very likely occurrence. In the long term, then, fixed wireless devices probably need to move out of unlicensed bands into their own dedicated bands, which might be above 20 GHz, in areas that are of less use to W-LANs and PANs because of the need for LOS propagation.

5.3.2.3 Interference to PANs

Because they move with people, PANs might expect interference from all possible sources. Individuals might be near W-LAN transmitters, near fixed wireless base stations, or actually using any number of different remote control devices. PANs also might experience interference from other PANs. There is little that can be done about this since there is little likelihood of PANs getting their own spectrum, and even if they did, they would still be interfered with by other PANs. Hence, PANs will need to be designed to overcome interference, which will make them less spectrally efficient, as discussed in Section 5.3.2.5.

5.3.2.4 Interference to remote control devices

Remote control devices normally work by sending a short pulse of energy. If there is excessive interference, the transmission will not be received and the user will normally have to move a little closer or adjust his or her position. The key effect of interference in this area is likely to be a reduction in range. Because most remote control devices are used in the home or away from buildings, most interference is likely to come from PANs, and given the very low power levels that are used here, this may not be unduly problematic.

5.3.2.5 Designing for interference

It is possible to design wireless communications systems to be more tolerant to interference. There are many ways of achieving this:

- Using CDMA, which is inherently robust against interference, although at the cost of reduced capacity;

‣ Using frequency sensing techniques that avoid frequencies on which interference is being received, which works as long as the device transmitting the interference is narrowband and stays on the same frequency;

‣ Using frequency hopping, resulting in any interference being problematic for only a limited number of hops, which tends to work only if the interference is limited to a few specific frequencies;

‣ Using error-correction coding to remove errors caused by interference, which can work but results in lower data rates as more redundancy is added to the signal;

‣ Using directional antennas to strengthen the wanted signal and weaken the others, which can be effective but will probably be difficult and expensive to implement anywhere but on cellular and fixed wireless base stations.

These techniques involve a number of key points:

‣ Most reduce the efficiency of use of the spectrum, resulting in lower transmitted bit rates than would otherwise be the case.

‣ Most rely on the source of the interference being relatively static. Once the source of the interference becomes a number of other "intelligent" transmitters all trying to reduce the interference their receiver is experiencing, some of these will techniques no longer work.

‣ None will work when there are high interference levels across all the different bands.

5.3.2.6 Summary of unlicensed issues

Unlicensed spectrum has become very attractive because it is free and it is available immediately, without the need for a long allocation process. Furthermore, many of the bands are available worldwide. However, the increased load that might be expected as W-LANs, PANs, and fixed wireless systems become widespread threatens to generate excessive interference in these bands. In particular, fixed wireless systems would seem vulnerable to interference and should probably not be used in these bands in the long term. PANs and W-LANs may be able to continue to use these bands but will need to be designed in such a way that they can overcome severe interference.

5.3.3 Future potential spectrum bands

Outside of the bands discussed above, there are other spectrum bands that may be of use in the future. Bands at 60 GHz and above are often discussed for future high data rate radio systems. The advantages of these bands are the enormous amount of spectrum available—often many gigahertz. The downside is the need for LOS propagation and the high device costs. The 60-GHz band itself is interesting because it is the oxygen absorption frequency, resulting in signals being rapidly attenuated. While this dramatically reduces range, it also facilitates very close reuse of the spectrum, resulting in massive capacity per link. To date, there has not been sufficient need for the capacity to make the cost of these links worthwhile, but this may change in the future.

At even higher bands lies the optical spectrum. During 2000 a number of proposals from companies such as Terabeam were put forward to use this band. These proposals were mostly for point-to-point optical laser links. The advantages of these links are the following.

▸ Potential for extremely high capacity, with early links offering up to 1-Gbps data rates, and a potential for higher rates;

▸ Use of unlicensed spectrum allowing immediate deployment with no regulatory issues;

▸ Laser technology is well advanced resulting in relatively low-cost equipment (compared to, for example, 60-GHz RF technology).

The disadvantages are listed as follows:

▸ LOS operation is essential;

▸ The link can be disrupted by fog or heavy rain and impaired by direct sunlight;

▸ The maximum range is typically of the order of 1 km.

On balance, the advantages are compelling; hence, there was much market activity in this area during 2000. It certainly seems plausible that optical equipment will take much of the market for wireless high-speed connections to businesses, but the equipment costs are typically too high for residential users. Many have predicted that optical systems will soon render LMDS solutions obsolete.

5.3.4 Ultrawide band (UWB)—a solution to the spectrum problem?

During 2000 a new technology was coming to the fore—that of UWB. The principle of UWB is to transmit short abrupt pulses of radio energy. Because these have sharp transitions in the time domain, they result in a very broad frequency spectrum. One of the key solutions proposed resulted in transmissions that spread over a band from 1 to 3 GHz. Much like CDMA, however, because the information is spread over a wide band, it is possible to transmit at a very low power level and use the processing gain resulting from the "excess" bandwidth usage to recover the signal from the noise. Hence, the principle of UWB is to transmit in bands already occupied by other transmissions but at such a low level that the signal is below the noise floor of the existing devices and thus does not impact their operation. This typically results in transmitted power levels on the order of 50 μW.

The resulting system has the characteristic of relatively high data rates—up to around 10 Mbps with the possibility of going higher as technology improves—but very short range, perhaps no more than 10m. This restricts its usage to PANs and W-LANs.

At present, UWB is illegal since it results in transmission in spectrum bands owned by others. Many regulatory bodies, in particular the FCC, were examining its use during 2000 and were expected to legalize it during 2001.

Assuming it is made legal, it is hard to see a leading role for UWB. Certainly, it could become the physical layer for PANs and W-LANs and might be attractive in the case that the spectrum that they utilize, generally unlicensed, became so congested that communications became problematic. As we have argued in Section 5.3.2, however, we do not expect this to be the case inside buildings. Outside of buildings, the range of UWB seems too short to make it a viable hot-spot W-LAN product.

It is hard to judge such an embryonic technology and easy to be skeptical of anything new. UWB may well have a role in the future. It is difficult, however, to see it solving the spectrum problem in general; at best it might overcome congestion in some unlicensed bands.

5.4 Finance

Regardless of whether all the constraints relating to spectrum, standards, and technical issues can be met, few organizations will move ahead with a project unless it meets their financial criteria. Organizations typically have the option of undertaking more projects than they have the resources to do, and so they select among projects on the basis of some financial measures. Typical measures that are used are detailed in many textbooks; for an introduction, see [1]. A brief guide follows.

▶ *Net present value (NPV).* This is the total of the discounted cash flows over the lifetime of the project, or over an agreed time horizon. Within the mobile telecommunications industry, it is typical to consider programs over a horizon of between five and 10 years. Discount factors are normally in the region of 15–20%.

▶ *Internal rate of return (IRR).* This is a measure equivalent to the interest that a bank would need to offer on money deposited in order for the return to be as good as the project considered. It is closely related to NPV (the IRR of a project is equal to the discount rate which would make the NPV zero). Within the mobile communications industry, IRRs in excess of 25% are considered necessary, with substantially higher IRRs often seen.

▶ *Payback period.* This is the length of time to pay back the initial investment. Until around 2000, payback periods were typically expected to be less than five years. With the increasingly large sums paid for third-generation licenses, however, it can only be assumed that these payback periods will increase up to 10 years.

▶ *Gross margin.* This is a measure used by manufacturers to model the difference between the selling price of goods and the cost of manufacturing those goods. A gross margin of, say, 40% implies that the cost of manufacture is 60% of the price at which the goods are sold. The remaining 40% is not pure profit since all overhead needs to be apportioned from this margin. Typical gross margin levels in the infrastructure arena are 40% but levels are lower on the handset side.

‣ Hybrid measures are often defined that, for example, include the NPV, the probability of the project being successful, and the strategic importance of the project to the company. These are all rated numerically and multiplied together to obtain a final weighted "score."

In order to illustrate these measures, a short example is presented in Table 5.3, which shows a hypothetical telecommunications operator building a new network and then gaining revenue as subscribers come onto the network.

Table 5.3 shows an operator with three outgoings and one revenue source. The outgoings are highest in the early years when the spectrum has to be paid for and the network is constructed. The revenue builds over the first few years as the number of subscribers grows and then declines in later years as the falling cost of airtime offsets the slowing rise in subscriber numbers. On an annual basis, the operator becomes profitable in 2004, but the debt is not repaid until 2007. To calculate the NPV, the annual profit is discounted by 1–15% (where 15% has been assumed to be the discount rate) for 2003, $(1-15\%)^2$ for 2004, and so on. These are

Table 5.3
Example of the Calculation of Financial Metrics

	Year								
	2002	**2003**	**2004**	**2005**	**2006**	**2007**	**2008**	**2009**	**2010**
Spectrum costs ($m)	1,250	0	0	0	0	0	0	0	0
Network costs ($m)	100	400	350	250	180	100	60	50	50
Operational costs ($m)	10	25	35	42	46	46	43	42	41
Revenue ($m)	0	200	480	720	880	1,000	920	840	720
Annual profit ($m)	−1,360	−225	95	428	654	854	817	748	629
Cumulative profit ($m)	−1,360	−1,585	−1,490	−1,062	−408	446	1,263	2,011	2,640
Discount factor	100%	85%	72%	61%	52%	44%	38%	32%	27%
Discounted profit	−1,360	−191	69	263	341	379	308	240	171
Discount rate =	15%								
Sum of discounted profit (NPV)		220	$m						
IIR		21%							
Payback period (year when cumulative profit is positive)	6th year (2007)								

then totaled. The IRR is the percentage discount rate that would result in the NPV being zero, but it is more readily calculated using functions built in to most spreadsheets.

A hybrid "score" might then be derived as follows. The operator may decide that the probability of success of this project was 75% and that on a rating of 1–5, where 5 was the most important, the strategic value of this project was 4. Then the score would be the NPV of $220 million × 75% × 4 = $660 million. This could then be compared with other projects also competing for investment.

A manufacturer or operator will use these metrics to rank multiple projects. They will then determine the available resources and, in principle, select the projects with the highest ranking until all the resources are consumed. In practice, few organizations can follow such a clear and logical path since other organizational issues often intervene, but nevertheless, successful organizations tend to act approximately in this manner.

The key implications of this are, first, that future projects will only be implemented if they meet appropriate financial thresholds, and, second, that even then they will only be implemented if they rank sufficiently high against all the other projects available to the organization. Simplistically, the most profitable technologies or services will be developed first, and the least profitable may never be developed. If we could predict which projects would be most profitable, we would have a good understanding of the drivers that will influence the decisions made by organizations in this area.

Historically, if we ranked the various forms of wireless communications, we might find that for manufacturers, the list would include the following (starting with the most profitable):

1. Cellular radio equipment;

2. Wireless LAN equipment;

3. Fixed wireless equipment.

The operators would be ranked as follows.

1. Cellular radio operation;

2. Fixed wireless operation.

Note that PANs appear in neither list because they do not yet exist, and wireless LAN equipment does not appear in the operator list because it is typically deployed by end users rather than operators. There are, however, a number of factors that lead to the conclusion that these lists might differ in the next 20 years. Cellular radio may well become less lucrative to the operator as a result of the dramatically increased license fees that are currently being paid for third-generation licenses and because of the ever-increasing competition in the cellular arena. If the operator is making less profit, then there will be increased pressure on the manufacturer to bear some of this loss through reduced equipment margins and revenue sharing schemes, also reducing the profitability to the manufacturer. Conversely, fixed wireless may be moving into a period of increased profitability as Internet usage increases and more users require and are prepared to pay for high-speed connections to the home or office. This demand, and the willingness to pay, will make the operation of fixed wireless networks more profitable and as a result, will likely make the manufacture of fixed wireless equipment more profitable. W-LAN and PAN volumes are likely to increase dramatically, but the pressure on price and the high level of competition will remain. Because these are open standards, maintaining a high margin will be difficult, and hence we might expect that these two areas will be less profitable. Looking into the future, we might conjecture that the list for the manufacturer would be the following.

1. Fixed wireless;

2. Cellular;

3. W-LAN equipment;

4. PAN equipment.

For the operator we might expect the list to reverse as follows.

1. Fixed wireless;

2. Cellular radio.

Such a change in priorities is profound, making a major difference to manufacturers and operators. Many operators and manufacturers today would be skeptical about such a change since the status quo has held for some 20 years for many in the industry, but the evidence points to a change of this sort. Because many will be skeptical and will retain their previous view of the industry, this will also be a time of change, with some manufacturers being eclipsed by newer entrants. For example, we might postulate that a long-time mobile manufacturer such as, say, Ericsson, might be late into the fixed wireless arena, whereas a new entrant such as Cisco might target this area exclusively looking for the new value that is being created. This might result in the value, and turnover, of these new entrants eclipsing the old companies. A similar story might hold true on the mobile side. This is often true when changes in the value proposition such as this take place. Existing players in the market fail to note these changes and are eclipsed by new entrants, often initially small companies that rapidly grow. Sometimes the old company takes over the new company and manages to integrate their model into the existing model, but sometimes this approach fails.

The discussion above suggests that at present we might expect that there are insufficient resources being utilized in the area of fixed wireless, and as a result, fixed wireless development might be slower than it otherwise could be. As time goes on, perhaps by 2005, this will swing to the other extreme, and fixed wireless investment will increase dramatically, perhaps at the expense of cellular. W-LAN and PAN manufacture will remain lower priority and might suffer as a result, especially with the large manufacturers. Smaller companies may fill this gap for the foreseeable future.

5.5 Environmental issues

Since around 1995 there have emerged a number of concerns related mainly to cellular systems: first, that the use of a mobile phone was detrimental to health, perhaps increasing the chance of cancer; and second, that the proliferation of masts was spoiling the landscape. Both of these have become significant problems for operators and might threaten to curtail the deployment of cellular systems in the future. This section looks at both of these issues.

5.5.1 Health concerns

It is well known that high levels of electromagnetic radiation are hazardous to one's health. Some RF engineers have had an above average number of cataracts in their eyes and other health problems.[2] Broadly, electromagnetic radiation, when absorbed by the body, warms parts of the body, particularly the eyes. Some have postulated that this warming might lead to the development of cancerous cells and other problems. It has long been widely believed, however, that for the power levels associated with mobile phones such health risks are highly unlikely. Others have postulated that there might be other effects not directly associated with heating that might be hazardous, but this has not been proven.

Since 1995 there have been many studies that have examined the effect of radiation on dummy heads or have conducted complex simulations to predict the warming effect of different kinds of mobile phones. The results of most of these studies have shown that mobile phones have minimal, if any, effect on the person using the phone. Proving conclusively, however, that mobile phones in no way affect health is almost impossible. It is very difficult to be absolutely certain that the radiation from a mobile phone might not be the catalyst for a cell that is nearly cancerous to finally become so. Therein lies the problem. Although nearly every study suggests that it is highly unlikely that there is any health hazard, none has proved conclusively that there is none.

In practice, health concerns seem to have had little effect on mobile phone penetration and usage since both continue to grow rapidly. It appears that for the general user, in the same manner as they accept that driving a car entails a small but nonzero risk of injury, they accept that the advantages of using a mobile phone outweigh the very small potential health risk.

Moving into the future, usage of mobile phones is likely to increase, as is the amount of electromagnetic radiation in the environment as fixed wireless systems, W-LANs, PANs, and other systems proliferate. Power transmission levels from the mobile device, however, are likely to be lower since the cell size decreases and so the distance to the base station decreases. Also, users may adopt techniques such as having the phone clipped to their belt and using BlueTooth to communicate with their headset such that the only transmissions near their head are very low power short-range transmissions.

Finally, it is possible that health concerns will decrease in the future as it becomes apparent that current users of cell phones, some of whom

---

2. The eyes are particularly vulnerable because they do not have their own blood supply, which can act to cool them.

had been using cell phones since 1980, are not any more prone to health problems than nonusers.

In summary, because it is extremely difficult to prove that mobile phones are not hazardous to one's health, health scares will continue for some time. These, however, do not seem to have overly concerned most users, and these concerns will decrease as power levels are reduced, mobile devices are moved away from the head, and it becomes apparent that long-term users of mobile phones are not suffering electromagnetic-related health problems.

An additional health issue that has been in the news in Europe in the late 1990s and the United States in 2000 is the use of a cell phone while driving. In much of Europe in 2000, it was illegal to use a handheld phone while driving, but not a hands-free version. In the United States both were legal, but many states were considering legislation in this area. The concern is that a driver using a cell phone would be distracted from the task of driving and thus more likely to cause an accident. Some countered that a driver talking on a cell phone was no different from one conversing with a passenger, and there does not appear to be any definitive work in this area yet. In the future, as will be discussed in Section 17.4, we expect all in-car use to become hands-free, which will overcome some of the problems. Others have gone further to postulate a filtering of calls to a driver depending on the current level of distractions that they face including "asking" them to pull over when an urgent call is incoming. While driving and using a cell phone is certainly a problem, it appears to be one that can be overcome by innovative ideas in the next 20 years.

5.5.2 Landscape issues

As four or more cellular companies seek to cover the landscape with more and more cells, the proliferation of cellular masts is somewhat inevitable. Cellular companies have adopted many approaches to try to reduce their impact on the environment. Some are hidden, for example, behind shutters in the roofs of churches; others are disguised, such as masts that are made to look like trees. Mast sharing is becoming more prevalent as masts owned by third parties are being used by many operators in the same area.

We may have reached a near peak in the number of large cell sites deployed in developed countries. Although third-generation operators will deploy networks, many of these will have limited wide-area coverage and will share masts used for second-generation systems. New cells will be smaller cells in cities and urban areas. Here, the environmental impact of masts is lower, and there are more places to hide masts. Furthermore, as cell sizes reduce, masts are no longer required and antennas can be placed on lampposts or in many other unobtrusive areas.

In summary, although landscape issues currently present serious problems to many operators, these problems may decrease rather than increase in the future as cell sizes reduce and deployment concentrates on urban areas. We might expect, then, that environmental issues, while they cannot be ignored, will not become a serious constraint on mobile communications developments in the next 20 years.

5.6   Billing, fraud, security, and numbering issues

This section looks at a number of remaining issues that act as constraints in today's wireless environment.

5.6.1   Billing

It is sometimes said that if you cannot bill a service, then you cannot provide it since you cannot collect revenue from the service and hence the business case will be disastrous. In today's networks billing is provided in a separate billing system—a computer connected to the switch. The switch generates call data records (CDRs), which provide information about each call made, and the billing system collects these and generates a bill for each subscriber. In principle, there is not much difficulty here—all that is required to introduce a new service is a new billing script, an addition to the billing software such that it is able to appropriately charge for the service. In practice, billing system development has proved to be a major block to the rapid introduction of new services in cellular networks over the last few years. This is broadly a logistical problem in that billing scripts can take some time to generate, and each change to the services offered or tariffs offered requires a change to the billing system. Only a limited amount of work can be completed in parallel due to the nature of programming billing systems, and the net result has been a large backlog of billing change requests for many operators. This has led to some setting criteria that new services can only be introduced if they do not require changes to the billing system.

Given that billing has proved to be one of the most critical factors for many operators in the past, it is reasonable to assume that the relatively slow rate of change of billing software will continue to be one of the key constraints on service development for some time. It is equally clear from the discussion above, however, that this need not be the case. Given that network operators perceive that a key competitive advantage is their capability to rapidly bring new services to market, the manufacturers of billing solutions will overcome the current problems by enabling rapid and parallel development of new billing scripts. This will be facilitated by the development of IP-based core networks where billing platforms can be connected into the network in a much simpler manner than the current switch-centric networks.

Other commercial developments in the future may also change the picture. For example, for services provided by third parties, these third parties may conduct their own billing, such as a monthly subscription fee to receive a daily newsletter—this fee being paid directly to the third party. Billing for the operator may even become simpler and be reduced to counting transmitted packets.

We note the current problems caused by billing systems and expect these to continue until around 2003–2005. By this time we expect that the combination of pressure to produce billing systems that can be more rapidly changed, the change in structure of the core network, and the change of commercial relationships to overcome these problems to a sufficient extent will result in billing no longer being a key delaying factor in the introduction of new services. It will still, however, absorb the time of many engineers and will remain a significant consideration in the introduction of new services.

5.6.2 Fraud and security

Fraud has been an enduring problem, particularly for mobile radio. Much of the impetus for the move from first- to second-generation mobile phones was the increasing fraud occurring because of the relative simplicity of overhearing mobile identification numbers and then making calls purporting to be from that phone. Digital cellular systems overcame many of these problems but also enabled a number of other fraud mechanisms such as the capability to steal a phone, set up a call forwarding mechanism from the phone before the theft was reported, and then utilize the forwarded path to make international calls at local rates. As the capabilities of wireless systems increase along with the range of services offered, the opportunities for fraud will similarly grow. As system designers close off known loopholes, fraudsters will use their ingenuity to devise new schemes to make money.

Perhaps there is nothing new here. Fraud has always been a battle between those trying to gain by some devious mechanism and those trying to protect their revenue. Typically, once a form of fraud is discovered, action is taken to defeat that particular form of fraud. In some cases, the action needs to be extreme, such as a move to a new generation of phone technology; in other cases a software patch on the switch is sufficient to prevent the fraud. For sure, fraud will continue to be an issue throughout the next 20 years, with many unexpected and ingenious tactics devised by the fraudsters. It seems highly unlikely, however, that fraud will stop the development of wireless technology and services. Indeed, it may even speed it by providing a strong incentive to introduce new technology.

Perhaps one potential area of concern is the likely introduction of personal agents. Personal agents are small, self-contained pieces of software. The idea is that in order to accomplish some task, a subscriber launches a personal agent into the network, and the agent is then "executed" by the network, returning the desired result to the end user. For example, a user may want to discover the lowest price hotel with certain facilities in a particular area. That user could log onto the Internet and search a number of sites to find the required information. Alternatively, users could just launch a personal agent that would visit all the appropriate Web sites, sort the information, and deliver the end result to the user. From the user's point of view this simplifies life, and from a network point of view it can, in many situations, reduce the amount of traffic over the air interface, normally the bottleneck in terms of bandwidth availability. Although the example quoted here is relatively simple, some have postulated that a user could have many hundreds of personal agents all working to filter and present information. These could even conduct transactions on the user's behalf without the user actually being aware this had happened, although the user would have approved the "principle" of the operation some time beforehand.

While undoubtedly useful, and perhaps essential as a mechanism for simplifying an increasingly complex environment, personal agents may present very fertile grounds for fraud due to their power, their autonomous nature, and their similarity in many ways to a virus. All sorts of fraud could be imagined, from the hotel that modifies the personal agent slightly to make their hotel the one returned to the end user, to personal agents that seriously disrupt network operations in ways similar to e-mail viruses such as the "I Love You" virus that appeared in 2000. In the same manner that the latter virus was overcome by blocking macros of a certain form, it may be that certain types of personal agents will be blocked.

Nevertheless, the skill of network designers will be tested, and it would seem that some network crashes will occur in the coming years.

5.6.3 Numbering

A shortage of telephone numbers has frequently caused problems to wireless network operators in the period 1980–2000. This was caused by increased requests for multiple telephone numbers—by 2000 a business user might have an office voice number, an office fax number, a mobile number, a home voice number, a home second line number for fax and e-mail, and a paging number. Interestingly, this proliferation of numbers was not something that users desired but was the only way that they could receive the services that they required. This increase in the quantity of phone numbers per subscriber led to a shortage of phone numbers in many countries, leading to a need to increase the total numbering pool by adding extra digits to all phone numbers in the country. Many countries had to completely renumber all phone numbers as frequently as three times in the period 1990–2000.

As discussed in more detail in Part IV, numbers, and particularly a proliferation of numbers, are not what the user wants. As can be seen by e-mail and Web addresses, users would prefer to have an alphabetical address of the form john.smith@some-ISP.com. Users would prefer a single address and have the network deal appropriately with incoming messages. Hence, we might predict that the trend toward an increasing amount of numbers will not continue much further, and perhaps by 2005 might start to decline as users have single unified addresses. Furthermore, these need not be centrally administered since they will be used for database look-up, which can then lead to network-specific addresses. Hence, our prediction is that numbering will relatively soon cease to become a constraint on the development and deployment of new services.

5.7 Summary of nontechnical constraints

The key constraints that have been identified in this chapter are described as follows:

> ‣ If no standard is developed in a particular area, this will provide a severe constraint on the future since standards are critical in enabling widespread wireless communications. However, as the need

for a standard becomes increasingly apparent, typically one or more bodies become prepared to undertake that standardization.

• Standards take roughly four to six years. Hence, this is a severe and important limit on the speed at which change can occur.

• In some cases, standards will be required that span a range of different industries, and the formation of appropriate groups to address this may be time-consuming and lead to the generation of competing standards. This is especially true of machine-to-machine communications and interworking between different networks.

• To gain the bandwidth required, it is necessary to move to increasingly higher frequencies, which leads to more problematic propagation. This will result in the spectrum available in large cells being little more than what is available today, but in small cells and fixed wireless systems much more spectrum will be available—typically on the order of 100 times more than today. The implication is that high data rate communications in large cells will be expensive but that much higher data rates will be achieved in small cells and fixed systems.

• Higher frequency operation requires more expensive devices. This will keep the operation of W-LAN and PAN systems below 10 GHz, but for fixed wireless systems, where a single transceiver is required per household (rather than the multiple transceivers for W-LAN and PANs), this will not be a great penalty.

• Unlicensed operation will become problematic as more devices use the same spectrum. In particular, fixed wireless systems will not be able to use unlicensed bands in the long term, although they may remain viable for W-LAN and PAN operation as long as these systems are designed to intelligently overcome interference.

• Manufacturers and operators will typically have more projects available to them than the resources they have to complete these projects. They will therefore select the most attractive. This will reduce the speed of implementation in some areas. This is likely to slow fixed wireless development until around 2005, and then after that W-LAN and PAN development might be expected to be impacted.

- Environmental issues such as health concerns and land-use concerns present serious problems to the operator now, but these are likely to lessen rather than increase as time goes on.

These constraints will be expanded upon in Part IV of this book where they will be used in the development of the road map. Chapter 6 considers the constraints imposed by organizations and strategic development models.

References

[1]   Webb, W., *The Complete Wireless Communications Professional*, Norwood, MA: Artech House, 1999.

[2]   Webb, W., *Introduction to Wireless Local Loop*, Second Edition, Norwood, MA: Artech House, 2000.

Contents

Organizations and strategy

"Never doubt that a small group of thoughtful citizens can change the world. Indeed, it's the only thing that ever has."

—*Margaret Mead (1901–1978)*

6.1 Introduction

Who develops the future? Rarely it is one single individual, although some individuals in history have made stunning changes. Equally, it is rare for it to be a significant percentage of the population. In areas such as mobile communications, the future is developed by the manufacturers and operators who are prepared to make the necessary investments to develop new technologies and standards and to deploy networks. If we look at the development of the original cellular phone, the concept was started within AT&T, one of the largest telecommunications organizations in the world. It was picked up by major manufacturers such as Motorola and Ericsson, companies with tens

139

of years of heritage in mobile communications of various descriptions. If we look at the development of GSM, although it is true that the core development was mostly pulled together by a team of around 10 people, these people were representing large organizations and well over 1,000 people were pulled into the standardization process at various points.

When considering the future, then, it is relevant to look at the companies who shaped wireless communications in the past and ask what actions they are likely to take in the future and what constraints they will be working under. It seems, based on aspects such as share valuation and movement of individuals, that new models are developing for the way that those involved in wireless communications are shaping the future and these models are developed and discussed in this chapter.

6.2   Today's wireless communications companies

6.2.1   The manufacturers

Mobile wireless communications today is dominated by a few key manufacturers: Motorola, Nortel, Ericsson, Lucent, Nokia, Siemens, Alcatel, and a few others. These are all massive companies, with valuations in excess of $10 billion and often $100 billion. They deliver huge infrastructure systems to operators and manufacture handsets in quantities of many millions. They spend billions of dollars per year on research and development and on attending and shaping standards. This list of manufacturers has changed little over the last 10–20 years because the barriers to entry are generally perceived to be enormous.

The perceived barriers to entry have not prevented some companies adopting more innovative models. Qualcomm has been stunningly successful in becoming an owner of key intellectual property rights (IPR), which they then license to these large manufacturers, taking a royalty payment on each unit sold. Other companies such as InterDigital have also followed this model, although with less success. Because of their ownership of key IPR and the likelihood that in the future they may develop and control other key IPR, Qualcomm has an ability to shape the industry vastly in excess of that which its turnover would immediately imply.

Other models are coming to the fore to allow new entrants into this marketplace. As explained below, many novel ideas are now developed in small companies, which are then typically acquired. In addition to the large wireless communications manufacturers, other companies, especially those who can leverage a rapidly climbing share price, can purchase these smaller companies. Companies such as Cisco are doing just that,[1] effectively making them a new entrant into the wireless communications arena. With the increasing capability to outsource manufacturing, large-scale manufacturing capabilities are no longer required. At present, a worldwide sales force is important and does represent a barrier to entry. Whether this will remain true into the future is less clear as many sales channels succumb to the Internet model and to electronic sales.

Although there are ways that new entrants can move into the mobile marketplace, these are difficult and it seems likely that any change over the next 20 years will be slow. Having said that, the pace of industry change in high technology areas is accelerating, so this is a difficult prediction to make. Probably one or two of the large manufacturers will make poor business decisions and will exit the business, or more likely, consolidate with another manufacturer. Probably one or two companies like Cisco will enter the wireless communications space and become key suppliers. Perhaps a few companies from the Asia Pacific region like Samsung or Sony will become key players in the marketplace.

6.2.2 The operators

Twenty years ago the only operators were the state-owned PTOs. Since then the picture has changed dramatically. Most countries have licensed between two and five mobile operators, of which one is typically the PTO. Since around 1997 we have seen these operators merge and maintain complex and multiple cross-holdings; this has resulted in mobile communications worldwide now being dominated by around 10 key operators such as Vodafone and France Telecom. Operators were originally started by companies with some history in telecommunications, but more recently utilities, banks, marketing companies, and others have all attempted to become cellular operators.

Entry into the cellular area is becoming increasingly difficult. License fees for new entrants are getting ever higher, and the new entrant faces well-established incumbents and an already fierce competitive environment. Only large companies, or consortia, can now realistically enter the cellular operator marketplace. The third-generation licenses granted by

---

1. The companies that Cisco has purchased are too numerous to list, and they typically purchase one every two weeks. Key purchases in the wireless space have included Clarity, a fixed wireless company, and JetCell, a GSM picocellular base station provider.

the end of 2000 have demonstrated the strength of the existing operators in winning the available licenses—in almost all cases existing operators have gained a 3G license.

In the fixed wireless arena the situation is very different. There are not many fixed wireless operators around the world today. Some are the PTOs, particularly in developing and Eastern European countries, using fixed wireless to speed their rollout of new lines. In developed countries the operators are typically new companies, such as Atlantic and the now bankrupt Ionica in the United Kingdom, and companies such as Fuzion in the United States. In fixed wireless the barriers to entry are much lower because there are fewer license fees, only limited competition, and no need to roll out a network nationwide. During 1999 there was the emergence of a number of pan-European fixed wireless operators such as Firstmark, and a number of U.S. operators such as Winstar started an international expansion, but this was still a developing process. In the future it will get harder for new operators to enter this marketplace because the competition becomes more intense as the spectrum becomes scarcer. Until around, say, 2005, however, we can expect to see a fair degree of volatility in the operators making up this market.

All is not straightforward in the role of the operator, however. The cellular operators are set on a course where "cellular rules the world." That is, they expect people to use cellular for all their mobile voice and data needs. As will be discussed in more detail in Chapter 17, there is some possibility that this may not be the case. While W-LANs will certainly dominate within the office and perhaps the home, they could also be deployed in public places, providing high data rate coverage in areas such as airports, shopping malls, and even throughout downtown areas. This model is already being explored in the United States by Metricom with their Ricochet product, which was offering service in Atlanta and San Diego in 2000. It has some key advantages, namely the higher data rates of W-LANs (in excess of 10 Mbps in some cases) and the fact that much of the W-LAN spectrum is unlicensed and hence free, unlike 3G spectrum for which cellular operators have paid significant fees. In the W-LAN model of the future, which we will explore in much more detail in the following chapters, subscribers use public W-LAN nodes whenever they are available and only resort to cellular where there is no W-LAN coverage. Although W-LAN coverage may only be 1% of cellular coverage, by carefully targeting busy areas it might be possible to capture as much as 50% of the traffic. The cellular operator may also be challenged in the home and workplace. At present, many cellular calls originate or terminate within buildings. If these calls transfer to corporate and home LANs the fall in traffic would be very significant.

If W-LANs do indeed play a major role in the future, cellular operators may decide either to "fight" them by attempting to prevent their deployment and offering competing services or they may decide to embrace them and become W-LAN operators in addition to their cellular networks. The former seems likely to result in some delay to W-LAN deployment but will unlikely prevent it completely. The option of fighting W-LAN deployment, however, would be a dangerous one for cellular operators since they may lose customers in the process and hence not be a route they are likely to follow for long.

Other trends also impact cellular operators. We are starting to see the first of the mobile virtual network operators (MVNOs), operators that do not own a network but lease network capacity from a conventional operator. They then manage the subscribers, acquiring them, delivering service, and billing them. In the future we might expect virtual operators to manage the subscriber relationship, allowing subscribers to access a range of different networks managed by conventional operators. MVNOs will emerge from strong brands such as banks, airlines, and other well-known service companies. They will extract some of the value from the subscriber relationship, reducing that gained by the conventional operators. Of course, the conventional operators no longer face the expense of marketing, service provisioning, and subscriber management.

So, in the future it seems likely that the role of the cellular operator will change. For some this will be a natural evolution. For others it will be traumatic and might involve acquisition or a reduction in influence. However it happens, it seems unlikely to have a major impact on the future—subscribers are able to rapidly move to more competitive companies and so operators have little capability to indulge in imposing their desires onto the future of communications.

6.2.3 The dot-com companies

The last five years have seen a change in the way that talented and innovative individuals perceive how they should develop new ideas. With the increasing availability of venture capital and the general buoyancy of the telecommunications environment, many now perceive that they can best be remunerated for innovative ideas by starting their own company and developing the idea to the stage that it is proven, with perhaps prototypes being available. At this point, generally, they will attempt to sell the company to one of the large manufacturers. This can result in large financial gain for the owners of the small company. The tendency toward this model has shifted the innovation out of large companies where innovators are less well rewarded (and often hampered by bureaucracy) and toward the smaller companies. The overall impact on the marketplace, however, is limited, because the large company tends to acquire the new technology in the end in any case, albeit at a higher cost than would have been the case had they developed it internally.

6.2.4 The role of the ecosystem

Worth a brief mention is a changing pattern in the way that companies, especially manufacturers, do business. A concept pioneered by Cisco is that of the "ecosystem." Just as in the natural world, this is the idea of a group of companies coming together for a particular project in a relationship that benefits all the different parties. In Cisco's case, this typically means a number of smaller companies providing services and equipment that the customers of Cisco may require, and that Cisco does not provide, coming together in a semiformal manner in order to make a consolidated offering to the customer group. On projects where there is limited benefit in collaboration, however, the same companies might be competitors. It is typically less formal than a consortia, but otherwise bears many similarities. It may well be becoming more common as companies concentrate on narrow parts of the value chain (for example, many "manufacturers" are outsourcing their assembly work) and hence require partnerships to deliver the complete service required by the customer.

This model has served Cisco and similar companies very well, and in an increasingly complex world where it is difficult for a single company to provide all pieces of the puzzle, this might well be a valid model for the future. It has its difficulties, especially when there are a number of large players in the ecosystem, but it might be expected that the companies that survive will be the ones that solve these problems first.

6.3 Strategy

According to Johnson and Scholes [1],

> Strategy is the direction and scope of an organization over the long term: ideally, which matches its resources to the changing environment, and in particular its markets, customers or clients so as to meet stakeholder expectations.

In its simplest terms, strategy is the set of plans that are put into place so that an organization can set its future direction based on the changes that it expects in its environment and marketplace. Simplistically, to set strategy a company would take a vision of the future that it believes in, determine how it can most profitably play a part in that vision of the future, and then understand the actions required to move from the current position to the vision of the future. In principle, then, strategy comes after forecasting.

Of course, without one or more companies developing a strategy based on a particular vision of the future, it is unlikely that this vision will come true. For example, if this book developed a vision that called for indoor LANs to be deployed in most homes, but no manufacturers decided to make inexpensive indoor LANs, then the vision itself would not become reality. Equally, if most manufacturers believed a different vision of the future, then they would tend to work in such a way as to develop product for that vision and convince operators of the need for that product. This would, in turn, result in their vision of the future being the one most likely to be realized.

Manufacturers and operators can, then, within reason, develop the future through the strategic decisions that they make. Sometimes, however, those decisions fail. For example, in 1996 many in the fixed wireless industry believed that the market was on the verge of dramatic expansion. Many manufacturers set in place strategies to capture a large share of the anticipated market. Operators were established to address the market. As the operators started to deploy, however, they found many problems. The equipment was too expensive; the radio spectrum they had to work with was problematic; and the competition was gaining in capabilities. The operators started to fail (e.g., Ionica in the United Kingdom), and this led to a decline in the orders for fixed wireless equipment and resulted in many manufacturers leaving the industry. Only in 2000 was the industry starting to reestablish itself. What is interesting in this particular story is the key role analysts played in influencing the industry. The analysts produced the forecasts that persuaded the manufacturers to enter the market that persuaded the operators, and so on. The role of analysts, then, and the way that they produce forecasts, is clearly important in how the future evolves.

What is clear, however, is that a prediction of the future that does not take into account the manner in which companies make strategic decisions may be inaccurate since certain restrictions as to, say, speed of strategic change may mean that companies simply do not react in the manner predicted. The remainder of this chapter explores the way in which wireless communication companies develop strategy and its impact on the future.

6.4   How is strategy developed?

There are many excellent industry texts on the development of strategy [1–3]. Broadly, they define the way that strategy should be developed:

> ‣ Analyze the situation by understanding the industry, the competition, and the current situation. Tools like strength-weaknesses-opportunities-threats (SWOT) and political-economic-social-technical (PEST) are often used as structured ways to analyze the current situation.

> ‣ Consider the possible strategies—again, generic models are available here such as low-cost producer, niche producer, and differentiated producer.

> ‣ Determine the best-fit strategy and implement.

There are many texts that devote hundreds of pages to all, or just a subset, of the above, and here we do not attempt to rival or repeat the information that they provide. Instead, we try to provide some insight into the manner in which strategy is developed in telecommunications companies in order to understand the impact that it will have on the future.

Strategy is the combination of all the decisions taken by a business as to the use of its resources. The aim of strategy is to maximize the return on those resources, although return may be measured in many ways including profit, turnover, and growth. Regardless of the measure, a good strategy is one that provides sustainable competitive advantage.

At the highest level, possible means of developing strategy could be divided into the following categories:

1. *Visionary strategic development* where a single individual, based on his or her assessment and understanding of the marketplace and his or her vision of the future, sets strategic direction;

2. *Consensus strategic development* where many individuals together form a strategy based on their combined wisdom.

Both have their strengths and weaknesses. Visionary strategic development tends to lead to dramatic successes when the visionary is correct but leads to dramatic failures when he or she is not. Consensus strategy tends to result in limited change in strategic direction, which tends to minimize failure but results in strategic drift and generally a gradual drop in market share. Many management advisors recommend visionary leadership during a time of change or loss of market share and then consensus strategy development during periods of market leadership and stability. We will return to this issue and the role of the visionary in shaping the future.

Some of the key strategic drivers of the telecommunications industry are listed as follows.

- For many areas, especially the mobile technologies and those areas where standards prevail, the cost of entering and partaking in the marketplace is high. For example, many large manufacturers spend 10% of revenue on research and development, amounting to nearly $1 billion per year in mobile wireless alone for companies such as Motorola.

- For infrastructure systems such as mobile and fixed wireless, sales are often $100 million to $5 billion and require a substantial sales staff, often the capability to deploy the system on behalf of the operator, and the provision of substantial funding often exceeding $10 billion for the total funding portfolio.

- For operators in most areas there is a need to obtain radio spectrum before they can deploy a service. Obtaining radio spectrum is time-consuming and often costly.

The strategic drivers listed above are especially true of the mobile cellular industry, and as a result there has been a trend to substantial consolidation, especially on the operator side where less than 10 operators now dominate the world scene. For industries such as in-building networks there are fewer barriers to entry for the following reasons.

- There is often not a standard or a much simpler standard that makes it possible for small companies to develop technology.

> • Because sales are often only to the owner of a single building or building complex, there is less need for expensive sales or for financing.

> • The technology itself is simpler because there is less mobility and thus less need for complex core networks, which again facilitates entry.

> • Many systems work in unlicensed spectrum, and as a result the operators do not need to obtain costly licenses.

Although small companies are able to play a role in shaping areas such as W-LANs and to a smaller extent fixed wireless systems, it will be the large companies that will have the biggest role in shaping the future. This is because the systems of the future, for which the requirements were discussed in Chapter 2, will be based on complex open standards, and only the large companies will have the finances and staff to attend and influence these standards and, perhaps, the overall vision and breadth of portfolio required to bring together all the different types of communications systems in the way that was envisioned in Chapter 3. Hence, it is worth looking a little at the manner in which these large companies develop their strategy.

As introduced above, strategy development is highly influenced by whether the company, or the sector, is run by a visionary leader or whether strategy is developed in a consensual manner. The visionary leader will form his or her own view of what the future should look like. Visionary leaders vary in style, but typically they will be individuals with a long exposure to the industry, and they will believe that they are able to predict where the industry is going based on what they know about customers, technology, and the competitors. There tend to be rather few visionary leaders in large telecommunications companies. It is not immediately obvious why this is the case—perhaps the complexity of the technology and the strategy development process and the diversity of different types of solutions make it difficult for one person to sufficiently grasp the issues to be sure of being correct. If they do exist, as far as their role on influencing the future is concerned, their actions tend to be difficult to predict, and thus they are likely to throw off track any logical determination of what the future might be.

For the consensual company it is common for strategic ideas and concepts to be developed in individual divisions. There may be someone within these divisions responsible for strategy who has a number of inputs:

- Discussions with existing customers enabling an understanding of changing customer requirements;

- Competitive intelligence information enabling an understanding of the actions and potential actions of the competitors;

- Forecast information both from external sources and based on discussions with customers and input from regional representatives;

- An understanding of the current product line, the planned road map, and the problems with that product;

- Attendance at conferences resulting in additional input information from consultants, analysts, and others involved in the industry.

These inputs will enable this individual to start to understand when their existing product and road map is insufficient to meet what they believe is the predicted market need. They may use common strategy tools to help them come to these conclusions, but typically their understanding of the market is already quite deep and they do not need SWOT analysis or formal identification of possible strategies.

Based on their assessment of the market, these individuals will make recommendations for particular projects to add to, or change, the road map. Possible proposals might include internally developing a new product, working with an external company in a partnership, or even acquiring another company. Each of these recommendations will have a certain cost associated with it, and the individual will typically calculate the cost and the potential revenue and determine the NPV of the project.

Proposals from a wide range of different divisions may then be passed to a portfolio management board. This board is responsible for the division of the scarce company resources among the inevitably numerous proposals for new programs. This board will typically assess a number of programs side by side, looking at the NPV, the resources required, and sometimes a concept such as the "strategic fit." The latter is difficult to define quantitatively. The aim here is to assess whether the project is more important than the NPV would suggest because it leverages increased sales in some other manner. For example, a proposal to develop dispatch capabilities in a cellular technology might not only result in increased sales, but if it were not undertaken, current sales might be lost as operators moved to a competitor's product. Based on a number of metrics, this portfolio management board will then recommend which projects the company should pursue.

Once started, projects are still not certain. During the life of the project, other projects may arise that will put additional strain on resources, and the portfolio management board may decide that a project that has already been started is less important than a new project that has just arisen. As a result, it may terminate the existing project.

Finally, and most importantly for the purposes of this book, there may be a sector-wide strategy team, not linked to any specific division. The purpose of this team is to look more widely at possible strategies that encompass a number of different divisions. It is here, for example, that strategies such as the linkage of cellular networks with in-building networks might be considered—strategies that each of the individual divisions would have less interest or skill in developing. This is where the breakthrough strategies are often developed. It is the relative strength of this group, which often has few resources and sometimes few senior managers, that determines whether a company tends to favor innovative projects that span different areas and attempts to lead the industry into the future.

It is tempting to believe that large companies can do almost anything—that they can develop whatever technology that they like and have limitless resources for building and selling complex systems. This is far from the truth. Large telecommunications companies have many different projects to consider and rarely have the resources to staff all the different projects. As a result, they are sometimes less able to develop technology in particular areas than smaller companies that concentrate on one product only and are able to get sufficient funding to develop that one product. This is particularly true of innovative new areas, which tend to be harder to justify in a large company that devotes much of its resource to maintaining and evolving its existing product. This is why large companies now often have to buy innovative new technology rather than developing it for themselves.

For small companies, the development of strategy is much simpler. They typically cannot afford to have individual strategists concentrating on specific areas. Instead, they will have been founded by a visionary who had a particular idea for a market segment or product. The company will then single-mindedly pursue that product, often with little regard for whether the market is actually changing. This single-mindedness tends to result in extreme outcomes. Typically, either the strategy was valid, in which case the technology will have been developed as fast as possible for a relatively low cost, or the strategy was not valid and the company will have developed a product that nobody requires and that will rapidly fold. Large companies cannot afford multiple project failures and thus tend to build in a lot more checks and balances to their projects.

6.5 A case study in strategic and organizational behavior

To illustrate some of these principles, it is interesting to look at a case study that encompasses many of these areas. This is the proposed development of the dual-mode DECT/GSM phone that was under consideration from 1995 to 1998. The development was ultimately unsuccessful, and we examine the reasons why here.

The basic concept sounded sensible. A phone would be developed that could operate on both GSM networks outside of the office and DECT networks within the office, allowing the user to have a single wireless phone in both these environments. The alternative was to have two phones, or to install in-building GSM systems. In 1995, however, the in-building GSM systems were in their infancy and were somewhat expensive. Both the GSM and the DECT standards were well developed and equipment was available for both, so there seemed little reason why this should not be a successful development.

In order for this initiative to be successful there were a number of requirements:

- An addition to the GSM standards was required so that phones could detect and register onto either the GSM or DECT networks as appropriate, and so that a single number could be used across both domains. Changes were required both in the mobile and in the network for the system to operate in the manner that the users required.

- The mobile phone needed to be sufficiently inexpensive and widely available so that there was little penalty in having a dual-mode phone compared to a single-mode phone.

‣ DECT systems needed to become widespread within offices in order for there to be a large market for the phone and for the infrastructure.

‣ Enough manufacturers needed to support the initiative so that users could buy from most of the major manufacturers.

None of these requirements seemed particularly problematic. The standardization work was relatively straightforward; it started around 1995, and concluded in 1997. Many chip-set developers showed how, in large volumes, the dual-mode phone might cost only $10 more than the single-mode GSM phone. DECT systems were slowly gaining acceptance in the office, although this was an area of some difficulty since most offices maintained fixed phone users. Initially, most of the key manufacturers showed interest in this proposal. By 1998, however, there was only one phone available from one manufacturer, which was bulky and expensive. By 2000 the initiative appeared to be completely dead. Examining why is interesting.

The key problem was that, in practice, very few manufacturers actually supported the initiative. Although many supported the standardization, which required only one person from each company, very few actually committed the resources to the development of the technology. This was because the manufacturers were faced at that point in time with a large number of projects requiring investment. These included wireless application protocol (WAP), prepaid phones, GPRS, EDGE, and early third-generation development work. They had limited resources and were not able to recruit skilled engineers because the cellular boom had created a worldwide shortage of skilled staff. Their existing customers, the cellular operators, were asking for features such as WAP but had less interest in dual-mode phones because it seemed to have limited benefit to them. This is because instead of users using the cellular phone in the office, with a GSM/DECT phone they would be using the office phone system, resulting in less potential revenue to the cellular operator. Only those cellular operators who were also fixed line operators (i.e., the PTOs) stood to gain from this approach. Hence, when the manufacturers looked at their internal business case for each of the different features that they could select, they typically found that the business case for the dual-mode phone was poor because only a subset of their customer group was interested in it. Worse, for most manufacturers, significant development work was required because they were not currently DECT manufacturers. Of the three largest manufacturers, only one had any DECT manufacturing capability. Hence, for business case reasons, few manufacturers supported this initiative. Only one moved ahead to develop the phone.

Once it became clear to the operators who were interested in this service that it would be supported by only a single manufacturer, they started to become concerned about high prices and monopoly supply. Slowly, the operators reduced their interest in this concept and support continued to ebb away from the program. Seeing this, the remaining manufacturer did not invest the development effort required to produce a fully integrated phone. This, of course, resulted in the operators further distancing themselves from the program and then, finally, all investment being stopped. The program was over.

The key lesson here is one of investment priorities and momentum. Intrinsically, there was nothing wrong with this concept and much to recommend it; there were simply not sufficient resources available within the manufacturing community for a program with a fair, rather than good, business case to proceed. As it became clear that support for the program would be limited, others started to move away, resulting in the early momentum rapidly dying. As soon as this was apparent, the program was effectively over. This example illustrates that not only the technology, standards, and spectrum need to be in place, there needs to be sufficient resources and a cross-industry momentum supporting the concept.

6.6   How is the future invented—do we control it?

It is an interesting discussion to ask how the future is actually developed and whether "we" control it. Clearly, the future is invented by humans since machines do not yet have the power of invention. There are so many individuals and companies working in different ways, however, that there is no single point of control on the direction in which we are headed. What control there actually is comes more from the structure of society, from the rewards that capitalism provides for different types of companies, from the availability of capital and intelligent labor, and other aspects of society. Society, then, a concept developed by man, in itself becomes a way to shape man's future. Robert Pirsig [4] has gone further than this to argue that societies are effectively living organisms higher up the evolutionary ladder than humans and dictate the actions and evolution of humans themselves.

Others have noted that telecommunications networks are increasingly becoming more intelligent. The development of concepts such as the software agent, a piece of software able to visit diverse networks and collect and filter information on behalf of its "owner" will mean that the actions of networks will often be difficult to predict.[2] As more intelligence is added to the edge of the network, any control over the actions of the network will fast pass from a single centralized point to multiple diverse points. Increasingly intelligent software written to cope with this lack of control will result in the network eventually becoming "intelligent" in its own right, able to some extent to control its own evolution.

Although there is clearly some truth in the above observations, few would accept that the future really is out of our control. Many controls do exist, including governmental bodies that can regulate networks, standards bodies where collections of individuals can discuss sensible views of the future, and ultimately consumers who can choose not to buy or use products that concern them. Networks can be shut down if they appear to be acting in inappropriate manners, although it would take a serious problem to force operators to shut down networks on which millions rely and that deliver important revenue to the operator.

6.7 Key issues for organizations and strategy

The key conclusions drawn in this chapter are described as follows.

- ▶ The key manufacturers involved in mobile communications will change somewhat over the next 20 years, but many of today's key players will remain, augmented by a few new entrants, typically coming from other, related industries.

- ▶ In the cellular arena there may be a turbulent period as the role of operators change, both with the emergence of MVNOs and the possible use of W-LANs to supplement cellular networks. In the fixed wireless arena there will be a very volatile period until around 2010 when a few global operators will have emerged.

---

2. Software agents were not in existence in 2000, and hence their likely impact was difficult to assess.

▸ Unless a vision of the future is shared by most manufacturers and operators, the strategic decisions needed to realize that vision will not be made.

▸ It is increasingly difficult for large companies to develop radical new ideas. These will be developed by smaller companies, which will in turn be acquired by larger companies.

▸ Some might postulate that the future is "out of our control" as networks become more intelligent and evolve in unanticipated manners. This would render prediction of the future almost impossible. Fortunately, few would share this belief.

This ends the examination of the different constraints that reduce the "speed" with which the future can happen. Part IV considers the views of a range of experts and utilizes the material in the first three parts to develop a road map for the next 20 years. First, though, Chapter 7 examines how successful a forecaster would have been if in 1980 he or she had predicted the next 20 years.

References

[1]   Johnson, G., and K. Scholes, *Exploring Corporate Strategy*, New York: Prentice Hall, 1993.

[2]   Thompson, A., and A. Strickland, *Strategy Formulation and Implementation*, Boston, MA: Irwin, 1992.

[3]   Thompson, A., and A. Strickland, *Strategic Management: Concepts and Cases*, Boston, MA: Irwin, 1993.

[4]   Pirsig, R., *Lila—An Enquiry into Morals*, London: Black Swan, 1991.

Views of the future

How a 20-year forecast might have looked in 1980

"And this certainly has to be the most historic phone call ever made."

—*Richard Nixon (1913–1994), talking to the Apollo 11 astronauts on the moon (July 20, 1969)*

7.1 A look back in history

There are many who will say that it is pointless to try to forecast 20 years ahead in the fast moving world of mobile radio. One way to understand the difficulties involved in making such a forecast is to go back 20 years to 1980 and pretend that we were making a forecast of the next 20 years. Of course, it is always rather tempting to "cheat" since we have perfect knowledge of what actually transpired, but we can utilize the same methodology that we will follow in this book to bring a little formality to the attempt.

If we go back to 1980, we see that there were mobile radio services in most European countries but serving only a very limited number of subscribers. Table 7.1 provides some details of the services and the maximum number of users that they could support. Generally, all the services rapidly rose to their maximum capacity and so this is also a reasonably good estimate of subscriber numbers.

These cellular systems had many flaws. They were often half-duplex, the calling party often had to have some preknowledge of what part of the country the called user was in, and calls had to be operator connected. Demand massively outstripped supply, however, and by 1980 there were advertisements in many national newspapers calling for a mobile phone line and prepared to pay the current line's owner very attractive sums to give up their line. The mobile phone systems were almost always operated by the PTO.

Other developments were afoot. In Chicago, AT&T was testing out the first trial cellular system in the 800-MHz frequency bands. This went live in 1977 and allowed an assessment of the quality of service, engineering design procedures, and level of demand. It used 10 base stations, each with an average range of four miles, and provided 136 voice channels. By 1980 the system was close to its capacity of 2,000 users (full capacity was finally reached in 1981). One of the key findings from the trial was that the service was very well accepted by the market with more than 80% of users being satisfied. Widespread licensing of cellular systems in the United States was still some two years off in 1980, but it was clear that such licensing would eventually occur.

In Scandinavia, the Nordic Mobile Telephone group was working on the Nordic Mobile Telephone (NMT) specifications. The NMT system

Table 7.1
Subscriber Numbers in 1980 in Europe

Country	Introduction of Service	Maximum Number of Users
Sweden	1955	20,000
Britain	1959	4,000
West Germany	1968	30,000
Finland	1970	35,000
Spain	1972	1,000
Austria	1972	10,000
Netherlands	1972	5,000
Italy	1973	6,000

Source: [1].

design was started in 1969, but no systems were deployed in 1980—the first systems went live in 1981. The systems were deployed in the 450-MHz frequency bands using 180 channels. Forecasts made in 1979 suggested that there would be a total of around 45,000 users in the entire Nordic region by 1991. This forecast was revised in 1985 to 175,000 and re-revised in 1988 to 400,000, but that was still too low for the 580,000 actual subscribers in 1991. Mobile phones at the time of introduction of the service were around $1,500 each.

Although we are now leaping ahead from 1980, it is worth pausing to note that the underforecasting by over an order of magnitude was a problem that beset cellular around the world in the period from 1985 to around 1997. Only after around 1997 did growth slow a little (at least in percentage terms) as the penetration started to approach saturation levels.

Turning to other fields in the years before 1980, the home computer was starting to make a debut. Although the IBM PC would not emerge until 1981, there were a number of home computers at low prices that were rapidly starting to penetrate homes. Moore's Law (see Chapter 4) was known and its impact was starting to become apparent.

In the fixed arena, the only viable fixed network was the PSTN, supporting data rates of 2.4 Kbps through the V.29 modem, although, of course, very few were dialing in since there were few computers, no portables, and no real e-mail service or other network capabilities.

7.2　How might an expert have predicted the next 20 years?

To try and make the prediction of 20 years forward from 1980, let us use mostly the same methodology that we propose to use in making our prediction from 2000. (We will not use the Delphi part of the analysis where experts are consulted since this would be somewhat difficult to do and probably of limited value.) Here we will utilize a much shorter version in the interests of space, because the main purpose of this chapter is to test the forecasting methodology rather than develop a comprehensive forecast.

In Chapter 2 we developed an argument to suggest that in the future, people might want any transmission of information between people, between machines, or between machines and people. We could then conceive of the ultimate wireless communications *device*—defined as a machine whose primary purpose is to enable wireless communications—as being one or multiple devices that enabled a person to:

- Communicate to any other person or groups of person at any time and regardless of the location of the individuals using any mode of communications available to humans but probably limited to video (with speech), speech alone, or text;

- Communication with any machine, potentially including machines not yet invented at the time the communicator was manufactured, in any meaningful manner for the human involving video, speech, and text.

We could conceive of the ultimate communications network as one that:

- Supports any number of the above devices;

- Allows any machine to communicate with any other machine in any meaningful manner.

This argument is not time-related—it would have been possible to derive this argument at any time after wireless communications became possible.[1] The key differences in putting forward this argument in 1980 and 2000 is not the argument itself but the constraints that apply to it—in 2000 the fact that everyone can have a cellular phone is more or less taken for granted; in 1980 this was far from the truth.

The key differences between predicting in 1980 and predicting in 2000 are described as follows.

- The constraints are clearly different—less can be expected in the capabilities of wireless communications. This is something we will examine in more detail in Section 7.2.1.

- Less can be envisaged. For example, with computer penetration being limited, it is hard to envisage the arrival of devices such as

---

1. Indeed, it would have been possible to derive it before wireless communications became possible, but it would have been difficult for most people to envision at such a time.

PDAs and to look ahead to a future where even toasters will have built in microprocessors.

▸ The competitive environment was less well developed, with typically a single monopoly telecommunications provider in each country; and as a result it was difficult to imagine a pace of change any more rapid than had been the case historically.

It is now worth taking a quick look at the constraints that applied in 1980, using the same format that we have looked at the constraints from 2000, but in less detail.

7.2.1 Constraints

7.2.1.1 Technical constraints

Of course, the overriding constraints were no different from those today, and Shannon's Laws were well known. Cellular systems were less efficient than they are today—in Chapter 4 we noted that third-generation systems are capable of around 30–50 "calls" per megahertz per cell. If we take a system such as the NMT system that was just about to go into service in 1980, we find that the speech channels were 25-kHz-wide, and that the reuse pattern was around 21 but each cell had three sectors, resulting in around six "calls" per megahertz per cell, almost an order of magnitude less than the capacities predicted today. Although much theory was in place, there was still much work that remained to be performed in order to improve this rate. Some were talking about using digital transmission, and a far-sighted person might have predicted the arrival of a digital cellular system during this period with perhaps a two to three times improvement in spectral efficiency. Hence, it might have been thought that the capacity would rise to perhaps a maximum of 15–20 "calls" per megahertz per cell over the 20-year period. The capacity of the system leads naturally to discussions of whether capacity is likely to be a restricting factor. Although all the forecasts made around 1980 for cellular penetration were around an order of magnitude too low, the astute forecaster would have been less concerned with what others were saying and more interested in questioning the fundamental drivers behind determining what the growth would actually be. Since all the cellular systems then in existence were operating at full capacity, he or she might well ask what the capacity of the next-generation systems would be and whether they too, would soon reach a limit.

It was well known by this time, as has been explained in Chapter 4, that the capacity is driven by the spectrum availability and the cell size. It was known that 450- and 900-MHz spectrums would be suitable candidates for cellular systems, although it was generally thought that above 1 GHz, propagation might be too difficult to viably deploy a cellular system. Indeed, it was not until around 1992 that serious consideration was given to using the 1.8-GHz spectrum as a cellular expansion band. So let us assume that the forecaster focused on systems below 1 GHz, allowing a maximum of perhaps 2–50 MHz of spectrum when both 450 and 900 MHz (or 800 MHz in the United States) were taken into account. With an efficiency in 1980 of six calls per megahertz per cell, this enables 300 simultaneous calls per cell in the 1980s and perhaps 900 in the 1990s when the predicted digital system emerged. Usage patterns would have been difficult to predict, but the forecaster might have looked at fixed line usage, understood that cellular was much more expensive, and reduced the cellular traffic levels to perhaps 10–20 mE during the busy hours.[2] This led to a maximum number of subscribers per cell of 20,000 in the 1980s and around 60,000 in the 1990s. The forecaster next had to understand how many cells there would be. From a coverage viewpoint, he or she might have assumed, based on available propagation data, that a few hundred would be required to cover a country the size of the United Kingdom. If we say, perhaps, 300, this leads to a total system capacity of around 6 million in the 1980s, rising to 18 million in the 1990s. In practice, because of the uneven distribution of the population, realistic maximum subscriber numbers would have been much lower, perhaps somewhere around one-half to one-third at, say, 2 million in the 1980s rising to 6 million in the 1990s. This would have been well in excess of the forecasts for subscriber numbers, leading the forecaster to conclude, at least initially, that cellular systems would have sufficient capacity.

---

2. The Erlang is the unit of traffic intensity, which is a measure of average occupancy of, for example, a telephone channel during a specified period of time, usually taken to be the busiest 60-minute period of the day, called the "busy hour." An intensity of 1 Erlang (1E) means continuous occupancy of the channel during the time period under consideration. Hence, an intensity of 10 milliErlangs (10 mE) means that the channel is in use for 1% of the time, or 36 seconds per hour. It is usual to determine the average minutes of use per subscriber during the busy hour, then divide by 0.06 to translate this figure into a busy hour traffic intensity expressed in milliErlangs.

7.2.1.2 Smaller cells

The forecaster would also have been aware of articles already published talking about microcells and a possible trend toward much higher density cellular architectures. Also, he or she would have seen the same constraints as discussed in Chapter 4, and would have dealt with them as follows:

▸ The base station equipment needs to be small and inexpensive. Base stations at the time were very large, often room-sized, and the miniaturization of these must have seemed a long way off. The forecaster may well have concluded that microcells would not be viable in the next 20 years.

▸ Linking the base station equipment back to the network needs to be simple and cost-effective. Backhaul was equally a major problem. The only real alternative was a leased line, which was generally expensive (although often the PTO was the only cellular operator and so this was an internal cost transfer). Again, this must have seemed a nearly insurmountable problem for microcells.

▸ The complexity of managing numerous base stations in such areas as frequency assignment needs to be overcome. Actually, the forecaster may not even have realized that this was going to become a very complex problem and so might not have been overly concerned about this.

▸ Handover and other mobility features become more critical in small cells as less time is spent in each cell. Again, the issues associated with handoff were not well understood, so the forecaster may not have realized the potential issues here.

In any case, the forecaster may well have concluded that small cells were unlikely to be viable within the next 20 years, and so the capacity would be capped at the levels outlined above.

7.2.1.3 Standards

There was already one standard being developed in the 1980s, the NMT standard, and it would have seemed reasonable to predict that other cellular standards would be developed. Given the activities of bodies such as CEPT, it would even have been reasonable to predict the eventual arrival of a European standard. In other areas of wireless communications, it is unlikely that the forecaster would even have considered their arrival, let alone the need for standards. Fixed wireless was not a concept that anyone was actively considering. Wireless LANs were not under consideration at a time when few even had LANs, and PANs were certainly not something anyone had thought about. As a result, standards probably did not seem likely to be a key issue in the 1980s.

7.2.1.4 Spectrum availability

Broadly, in 1980, spectrum availability would not have appeared to be a key issue. Given the expectation in subscriber and traffic levels, as has already been demonstrated, there was apparently ample spectrum. What transpired was somewhat different, and by 1990 many countries were looking for ways to increase the spectrum for cellular communications including sharing with the military and clearing additional spectrum in new bands.

7.2.1.5 Financial constraints

These would also generally not have been seen as problematic in 1980. Most cellular operators were PTOs with deep pockets. Although the finance community had not realized the potential returns from cellular, there were many large companies willing to invest in this area.

7.2.1.6 Environmental issues

Health concerns and land-use issues were also not problematic in 1980. At that time there was no suggestion that there would be any health issue. Also, the number of cell sites predicted was much lower than actually transpired, and hence the number of masts causing landscaping issues was limited.

7.2.1.7 Organizations and strategy

In the 1980s the only organizations involved in telecommunications were the PTOs. They had been the monopoly suppliers for 80 years or more, and it would have been hard to predict that this was about to change in the next 20 years. Certainly, had someone predicted that the cellular operators, then nonexistent, would have a larger market value than the PTOs within the next 20 years, this would have seemed rather ludicrous. In the United Kingdom in 1980 there was some preliminary discussion about having a competitor to BT when cellular licenses were distributed, which took place in 1982, but it was far from certain that this would occur or that any competitor could seriously challenge BT. PTOs were known for being slow to change, and so the forecaster would have assumed that cellular systems would be rolled out slowly and that prices would only fall slowly over time.

On the manufacturer side there was a thriving community including companies such as Motorola, Ericsson, Siemens, Ferranti, Marconi, and others, so it could have been assumed that competition would drive prices down. There had already been many examples of consumer electronics prices falling over time, and it would have been safe to predict that cellular terminals (handsets were not available at the time) would follow a similar trend.

7.2.2 Requirements

Having now considered the constraints, let us again look at the requirements and try to determine how a forecaster in 1980 would have dealt with them.

> Communicate to any other person or groups of people at any time and regardless of the location of the individuals, using any mode of communications available to humans but probably limited to video (with speech), speech alone, or text.

With regards to the ubiquity of communications, the forecaster might have expected that most of a country the size of the United Kingdom could eventually be covered. He or she might well have thought that rural areas would not be covered for some time and expected that communications within buildings would not be ubiquitous. It was known that radio signals would penetrate buildings but that some rooms within the interior would have poor coverage. Communications from planes or in remote locations would have seemed unlikely.

With regards to the mode of communications, speech was clearly going to be the main mode and had already been demonstrated. Text would have seemed somewhat pointless, given the very poor display technology. The forecaster might have supposed that text would not arrive until digital systems in the 1990s and even then would have limited application. Video communications had often been mentioned in the research establishments but would have still seemed a long way off. Video compression was in its infancy and the bandwidth required for video would have seemed completely incompatible with cellular communications. Hence, the observer would have assumed that almost all communications would remain voice-based in the next 20 years.

Communication with any machine, potentially including machines
not yet invented at the time the communicator was manufactured,
in any meaningful manner for the human involving video, speech,
and text.

Although the forecaster might have thought this one up as a general
requirement, he or she would quickly have discounted it, mostly because
there would not seem to be any point to it. The thought of the refrigerator talking to a PDA in order to update the shopping list would have
seemed fanciful at a time where computing devices were relatively large
and relatively expensive. Not even the most forward-looking observers
seemed to be proposing such schemes. Perhaps the forecaster might have
thought that there would be a time when we would all have computers
and we might want to access the computing device from our mobile
phone. They would not have really been able to predict the applications
that would be enabled by such connectivity but perhaps considered issues
such as home control where an individual could dial from his or her
phone to turn on the heating in the home.

7.2.3 Subscriber numbers and penetration
This brings us back to subscriber penetration levels. The forecaster has
already determined that by the 1990s, cellular systems with capacities of
up to 18 million subscribers for a country the size of the United Kingdom
might exist. The U.K. population is around 60 million, so this represents a
penetration level of 30%. All the predictions, however, pointed to much
lower numbers than this, perhaps 1–2 million. These predictions would
have been supported by charts showing the take-up of technologies, such
as the one shown in Figure 7.1.

Although many of the technologies on the chart shown in Figure 7.1
were still to be invented in 1980, the VCR had been in existence for five
years and had still to reach any significant penetration. Cable TV, and
indeed most forms of TV, had taken around 15 years to reach 20% penetration. Cellular, somehow, seemed more of a business-related tool, and
so it was expected to probably only reach 5–10% penetration.

Much of this misconception is now clearly based on the history of
cellular at that time. Subscriber numbers had been low primarily because
system capacity was low, but the visible impact of this was high prices
and use by an elite. It was hard to extrapolate that one day this service,
which seemed almost magical and existed mostly in Rolls Royce and

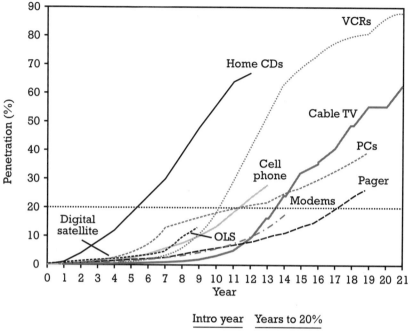

	Intro year	Years to 20%
VCRs (1)	1975	10
Home CDs	1985	5
Cable TV	1954	14
PCs (2)	1977	11
Pager	1978	16
Cell phone	1983	11
OLS	1988	>10
Modems	1982	~15
Digital satellite	1994	?

Figure 7.1 Adoption rates of a range of different technologies.

Mercedes cars, would be a tool for the masses. Most forecasters would have accepted this and gone with the prevailing opinion that maximum penetration levels might reach around 5% of the population at 3 million in the United Kingdom by 2000. Perhaps a visionary might have questioned why the potential capacity of 18 million would not be reached but quickly would have backed away when presented with the evidence that the introduction of such new technology into society was necessarily a slow process, which had been the experience of the past.

7.2.4 Summary of what a forecaster might have predicted

In summary, using the methodology outlined above, the forecaster might have predicted that analog cellular systems would be installed during the 1980s, perhaps being replaced during the 1990s with digital systems, with coverage of most of a country such as the United Kingdom and with more than enough capacity to meet requirements. The system would be comprised of large cells, with perhaps 300 covering a country like the United Kingdom. It would be unlikely that microcells or in-building cells would be deployed. The forecaster would have predicted that the systems would be essentially speech-based systems with no video capability and with limited text capability. Furthermore, the forecaster would not have assumed any real interaction with machines, except perhaps the capability to dial up to the home computer by the end of the 1990s in order to perhaps interact with an intelligent house. The forecaster would have foreseen penetration rates of perhaps 5% or less. Indeed, the key prediction would have simply been the development of a mobile communications system that businessmen could use with a handheld device within the next 20 years.

7.3 What actually transpired?

We do not need to dwell too long on what actually transpired since it is part of the everyday lives of most people in the developed world. For those who are interested in reading in detail about how the cellular business developed from 1980 to 2000, [1] is an excellent source of information. In outline, analog systems were deployed in many countries from 1980 to around 1988, with cellular penetration levels reaching around 2% by 1988 in most developed countries. By 1988, standardization on second-generation digital cellular systems was underway, namely GSM in Europe and TDMA and CDMA in the United States. GSM was introduced in 1992 and subscriber numbers continued to grow very rapidly, reaching penetration levels of 30–60% in developed countries around the world by 2000.

For most of this time the cellular service remained a voice-only service. GSM provided short messages, using the short message service (SMS), but few used this; only by 1999 was this starting to be more than a curiosity. GSM also provided data capabilities. These started to become available on networks in 1994, and at first the data service was used by only a few traveling businesspersons as a way to link their laptops into the office network and remotely retrieve e-mail and other data. Even by 2000, the data service made up less than 2% of the total call volume since the difficulties and cost of using data remained high. Videophones remained a research topic, appearing only as concept handsets for third-generation systems. As mentioned in Chapter 4, by 1998 small base stations had started to appear and in-building coverage was becoming an important topic, albeit one where there was limited commercial activity. Second-generation systems continued to evolve, and by 2000 packet data services were just about to be launched using GPRS and other similar standards, but these were not expected to become widespread until 2001 or 2002.

In other areas reasonable progress was made. W-LANs were implemented in some locations and proved themselves reliable and useful, if somewhat expensive and relatively slow. Satellite systems arrived with the advent of Iridium to commercial service in 1998, followed very rapidly by its demise into bankruptcy in 2000, leaving much doubt as to whether mobile satellite systems had a realistic future. Paging and two-way radio systems made reasonable progress, although they never really became part of the mainstream communications network and tended to be confined to particular professional groups.

Now we need to compare this with the hypothetical forecast that an expert would have made in 1980. On a point-by-point basis, such a forecast would contain the following information.

▸ Analog cellular systems would be installed during the 1980s, perhaps being replaced during the 1990s with digital systems. This is precisely what happened.

▸ There would be coverage of most of a country, such as the United Kingdom, and with more than enough capacity to meet requirements. The coverage prediction was correct, but there proved to be insufficient capacity in the basic network design and so many capacity enhancement techniques were adopted, including frequency hopping and microcell.

▸ The system would be comprised of large cells, with perhaps 300 covering a country the size of the United Kingdom. In fact, each operator put in many more, with some having more than 2,000 cells and growing rapidly.

▶ It would be unlikely that microcells or in-building cells would be deployed. These were just being deployed from about 1998 onward.

▶ The systems essentially would be speech-based systems with no video capability and with limited text capability. This is what transpired.

▶ There would be no real interaction with machines, except perhaps the capability to dial up to the home computer by the end of the 1990s in order to interact with an intelligent house. Slightly more machine interaction occurred, mainly with computers, but broadly this premise was valid.

▶ There would be penetration rates of perhaps 5% or less. This was dramatically in error, with penetration rates between 30% and 60% by 2000.

What is striking is how accurate the prediction was in almost every respect, particularly in terms of the services offered. In thinking of the last 20 years, most would point to the single key error in the forecast: the huge difference between expected and actual penetration.

7.4   How significant are the differences?

Examining the differences in the forecast in more detail, we can understand how significant the errors would be for a range of different interested parties such as operators, manufacturers, and users. The key differences are listed below.

▶ *Prediction of sufficient capacity.* The prediction that capacity would be sufficient when in fact it was not had major implications, particularly for operators, and to some extent for users who found high blockage rates. Operators who built networks on the premise of sufficient capacity soon found that they had to modify their networks, adding sectored cells, adding additional cells, replanning frequencies, and conducting other expensive changes. This, however, was a nice problem to have. The cost of reengineering the network was soon paid for by the higher number of subscribers, each providing additional revenue. For the manufacturers, more effort needed to be spent in developing solutions to the capacity problems, which hastened developments of technology such as frequency hopping.

‣ *Three hundred cells would be sufficient to cover the United Kingdom.* This prediction was probably true in strict terms: More cells were required for capacity rather than coverage. The issues, then, were very similar to those developed above.

‣ *Penetration rates would reach 5%.* This had a major impact on everyone in the cellular industry. The operators had to reengineer their networks, change their models of selling in order to sell to private as well as business users, change their tariffs, provide handset subsidy, and modify other ways that they conducted business. As before, these were nice problems to have, and their profits and value rose much more rapidly with the increased penetration than would have been the case had penetration rates been as expected. For manufacturers, this led to increased sales, again leading to higher valuations. For the end users, this enabled those who previously would not have considered owning a cellular phone to become a cellular phone owner and user. By 2000, fixed-line operators were starting to see their voice traffic levels fall as people used cellular phones in preference to fixed phones, a substitution that would have been difficult to envisage in 1980.

So how significant are these differences between the forecast and reality? To some extent that depends on the point of view. For the fixed-line operator, they are significant in a negative manner. For the cellular operator, they are significant in a positive manner. For the end user, they are in general significant in a positive manner. The real issue is whether the user of such a forecast would find the forecast made in the 1980s useful. To understand this, we will take a look at the viewpoint of each of the interested parties.

‣ *Cellular operator.* In general, the cellular operator would find this to be an extremely useful forecast. It predicted with accuracy the fact that there would be two generations of cellular systems, that there would not be microcells, that the main service would be voice, and that there would be a limited need for data. Based on this, the operator would be able to determine that it was worthwhile to deploy a cellular network, would know the lifetime of the network, would know the sort of services that would be deployed, and much more. Most operators today would be delighted to believe that they could predict the next 20 years with similar levels of accuracy. On the downside, the operator would clearly have liked an accurate demand forecast. In practice, the operators were able to modify their networks in a relatively painless fashion to add additional capacity. They did not need to remove the network and deploy a different technology, nor did they need to recall subscriber terminals or take any other drastic action. So, in summary, the cellular operator would have been well pleased.

▸ *Equipment manufacturer.* The position of the equipment manufacturer is very similar to that of the cellular operator. They would have been delighted to understand which technology to develop and the life span of that technology. They would also have been pleased to know that they did not have to develop data solutions or to be overly concerned about advanced services such as videophone. The error in the size of the market was less significant for the manufacturer—the original market estimates were sufficiently large to make market entry worthwhile, the larger than expected penetration just meant that factory capacity had to be increased, but, of course, profitability also was increased.

▸ *End user.* For the business user, the forecast was precise. It provided them with information as to the type of cellular network that they would use and the features that would be provided. They needed to know nothing more to be able to plan all wireless communication aspects of their business. For the private user, the 1980 forecast would have suggested to them that mobile communications would be out of their reach for the next 20 years. In practice, 15 years proved to be a more accurate time scale, and so many were pleasantly surprised that they were able to own a cellular phone before they initially thought that they could.

In summary, for most, the errors were not particularly significant in that few, if any, plans of action that were set out in 1980 would have been materially affected by what actually transpired. In reality, most things turned out better than originally expected, but not differently. We can conclude, then, that the forecast was successful and that we might have reasonable confidence in adopting the same methodology for the next 20 years.

7.5 What lessons can be learned and what is different now?

The key error in the forecast was that penetration was much higher than expected. This lesson has now been well learned by those in the cellular industry. Forecasts made five years ago tend to be fairly accurate. Perhaps the key error that might be made today is to overforecast. As mentioned in Part III, the growth that has been experienced in recent years cannot continue, at least in terms of subscriber numbers, because soon penetration will be at a level where virtually the entire population has a mobile phone. In 1980 the industry was in its infancy; in 2000 it had become a mature industry, which is well understood and for which there are 20 years of statistics. We might therefore expect that forecasting subscriber numbers would not be problematic for the next 20 years. Forecasting usage levels may be problematic, but not subscriber numbers.

Many would argue that forecasting 20 years ago was a much simpler task than forecasting today because today's environment is so much more complicated and less stable. There have been a number of paradigm shifts that make forecasting the future much harder. These include the Internet, which is changing much of society, the proliferation of PDAs like the PalmPilot, the development of PANs, the apparently increasing pace of technological change, and more. Indeed, Figure 7.1 shows that devices or systems introduced today tend to reach high levels of penetration much faster than 20 years ago. This points to a faster-changing society, which, in principle, would be one that is more difficult to forecast.

This is all true, but the fact remains that 20 years ago mobile radio also represented a paradigm change whereas today we understand mobile radio and the manner in which it will be used; so on balance it would appear that the forecasting difficulties are probably approximately equal, but the issues that will be important and the likely errors are different.

In addition to understanding subscriber numbers, we also have a good understanding of the speed of introduction of new generations of cellular technologies. Table 7.2 shows the dates of introduction of

Table 7.2
Prediction of the Introduction of New Generations of Cellular Technology

	Date of Introduction	Time from Previous Generation	Key Features
First-generation (analog)	1982	N/A	First cellular system
Second-generation (digital)	1992	10 years	Digital; provides data; increased capacity
Third-generation	2002	10 years	Higher data rates; additional capacity increases
Fourth-generation	2012	10 years	Additional enhancements in capacity and data rates; more services; integration with all other forms of communications
Fifth-generation	2022	10 years	—

different generations of cellular equipment, the time between them, and the likely future development of new generations.

What is clear is that there has been almost exactly 10 years between each generation of cellular technology. This is not accidental—it generally takes this long for technology to develop, for standards to be developed, and most importantly for operators to have gained sufficient payback on their existing system that they are able to contemplate expenditure on a new system. We already understand what a third-generation system will look like and will do. We can expect, in the period that we are forecasting, that there will be one additional generation of cellular technology, although we will need to derive what additional benefits this might bring over the third generation. Hence, we can expect cellular systems to be reasonably stable and predictable.

As has already been discussed, the difficulty arises in forecasting new areas, the areas of fixed wireless, W-LANs, and PANs, and in understanding machine communications. To some extent, these are very similar problems to those that faced cellular forecasters in the 1980s. Key questions include whether the service will survive, at what speed the service will penetrate the public, and what the utilization will be. With a much greater understanding of the speed at which new devices can penetrate the public, it seems unlikely that this will be underforecast in the way that cellular was. An overforecast is perhaps more likely, as happened with Iridium and with many recent Internet companies.

In summary, it would appear that a 20-year forecast in the area of wireless communications could be of substantial value to all the key players and that the methodology used here appears to be one that generates as good results as could be expected. Hence, it is worth proceeding to the final chapters of this book to learn and understand the predictions. However, we probably can learn little more from the prediction made in 1980. The lessons concerning penetration are well understood, and the fact that the forecast was accurate in other areas, such as technological development, is no indication that it will also be accurate in the future given the changing paradigm.

7.6 What the experts say—the next seven chapters

In principle, if everyone who would have an influence on the development of wireless communications over the next 20 years was consulted, and their views appropriately circulated, prioritized, and weighted by their influence, as good a prediction as could be made at this point would be achieved. Of course, those consulted might change their views over time, so there is no guarantee that it would be correct.

In Chapters 8–14, a number of prominent and important individuals in the world of wireless communications have presented their views of the future. After their views are presented, their diversity of opinion will be considered and discussed and then incorporated into the future road map developed in Chapters 15–17.

Reference

[1]   Garrard, G., *Cellular Communications: Worldwide Market Developments*, Norwood, MA: Artech House, 1997.

CHAPTER

8

Contents

Forecasting the future: a technologist's view

Mark Birchler, Larry Marturano, and Frank Yester

8.1 Introduction

The rapid changes in technology and the current era of major shifts in the wireless world make looking into the future extremely difficult. Trying to understand the forces that will shape the future, however, can be very useful. It can lead to a better understanding of future technology needs and future business directions.

When we were first asked to write a section of this book, we spent some time discussing the approach we would take. Being technologists, we wanted to consider some of the current and future technologies and some of their limitations. However, we realized that a good technology will not necessarily have a significant impact on the future. Its ability to affect the future will be determined by how widely it is deployed

179

and how successful it is in the marketplace. The market success of a technology is due to many complex, interrelated factors such as whether it solves a real user need, its acceptability to the user, whether it can be provided at the right price, and side effects of the product's use. These are further influenced by the regulatory environment and by social and political factors. Many things that are technologically possible never happen because of these other factors.

So looking at the future while considering only technology would give a very distorted view. Therefore, even though our background is in technology, we decided to try to address a broader set of factors. Obviously, this approach has its own problems. The number of issues to balance is much higher and their relationships are complex. In addition, our understanding of these other factors is more limited. We felt, however, that their integration was more important than a detailed discussion of technology in isolation. Our approach therefore considers what technologies could be available in 20 years and filters these with a consideration of the other factors to project what technologies will have a major impact on the future through their success in the market. We therefore ask your indulgence if we miss some key issues in these other areas or fail to emphasize their importance.

We also decided to try to identify the technologies that are expected to have the largest direct effect on people's lives. There are many important technologies that will be significant in some limited segment of the market, but because of limitations of time and space, we decided to focus on a few that will have the broadest and most visible impact on the future. Finally, it should be noted that since our approach considers successful deployment of technology, it should be recognized that research, standards, product development, and initial deployment will all need to occur prior to the indicated time frames.

8.2 Key factors

We believe that the wireless industry will migrate toward the provision of integrated services in which voice is simply one of many potential modes of communication. With the advent of BlueTooth technology, wireless handsets will quickly evolve to provide many different services including voice communication, messaging, entertainment, shopping, personal, and organization. Wireless voice communication will continue to consume ever-larger pieces of the total minutes of use pie at the expense of wired, but we expect the other services to experience rapid growth and dominate over the long term.

Some of the key questions that we need to address in order to understand the future are listed as follows.

- Will the home become a major incubator for new wireless devices and services?

- To what extent will entertainment products utilize wireless as a means of customer connectivity and distribution?

- Will the automobile continue migrating toward the "LAN on wheels," and to what extent will wireless be utilized to deliver content and services?

- To what extent will business users lead the ubiquitous delivery of services over W-LAN and wide area wireless systems?

- How fast will the market for services other than voice grow?

- To what extent will the market embrace wireless Internet-based services?

- How will the bundling of wired, wireless, and long distance impact the evolution of wireless systems?

- How fast will the wireless Internet bury itself in every piece of our environment, creating an interconnected intelligent landscape built with a myriad of communicating objects?

- Will Internet businesses gain control of the high value applications and services and make operators commodity transport providers?

- Will the earlier introduction of PANs limit the success of wireless LANs?

In the following sections, we hope to provide a context for further thought and discussion of some of these key questions.

8.3 Applications

8.3.1 Background

Applications are ultimately what the customer is paying for; therefore, they are a primary driver in the future of wireless. Manufacturers and operators can introduce systems, but if the users don't want the applications and services that these systems enable, they will not be viable.

When cellular radio was first introduced, it had one killer application—voice communication. The system merely took an application with which users were intimately familiar and added mobility. This simple concept first penetrated the business market where the cost of the service could be justified on the basis of higher productivity. It then began to penetrate the consumer market as costs came down and has now achieved penetration rates in many countries in the 30–60% range. Penetration of cellular use is continuing both in terms of number of customers and minutes of use per customer with wireless continuing to take minutes of use from wireline. This simple idea of providing mobility for a wired service created a major new industry.

Wireless operators, however, have been continually challenged to maintain their average revenue per user and to keep the rate of churn down. Operators are now seriously looking at new applications and services as a way to maintain revenue growth.

One approach we can use to get an idea of what these future wireless applications might be is to take a look at current or projected wired applications that we can make mobile. We can augment this by looking at applications that are unique to mobile or where mobility adds additional value as a second step.

Clearly, the home space is a fertile ground for identifying these new applications. There is a lot of emphasis currently being placed on the home market because of its size. People have clearly indicated a willingness to pay for some advanced applications and services in the home. The home space will encompass both business and consumer applications and services. As already seen with the Internet, particularly in the United States, the home market has driven a wide array of new applications and services for both business and consumer uses. With the growth of cable and DSL, we can expect a much larger segment of users to have access to much higher bandwidths. The higher bandwidths that are either currently, or soon to be, available in the home will drive user recognition and acceptance of new applications. With the continuing growth of telecommuting and work at home, major corporations will continue to expand their networks into the home with the goal of increasing employee productivity. Let's briefly look at the expected evolution of applications in the home.

8.3.2 Evolution of home applications

In projecting the future, there are many views, and it's not always easy to tell the plausible from the wishful. Everything is not equally likely, however, and using an understanding of the market and technology, we can at least note some of the things that are more likely and expose some that are blatantly unlikely. A book that takes this approach and makes sense of what we've seen in the development of another infrastructure-driven business (cellular) is *Residential Broadband* by Kim Maxwell [1]. The following information is primarily derived from that book.

The book's premise is that broadband two-way digital service to the home at high rates will evolve over time. It is unlikely to happen rapidly given the major changes that are required in home access networks and the supporting infrastructure due to the significant investment required. Therefore, the move to higher bandwidths in the home will occur in steps. For instance with DSL, it is possible to support dedicated 6 Mbps to the home in most areas by the deployment of equipment in the central office. To get to 26 Mbps, however, requires the construction of remote terminal sites within 9,000 feet of the homes they are to serve. These sites terminate the copper wires to the home and transition the connection to fiber. The next logical step would be fiber to the home, another costly infrastructure upgrade. New applications must be introduced and deployed to generate the capital required for this massive upgrade. Based on these types of considerations and past deployments, Maxwell's book develops a timeline for the deployment of broadband rates and the applications that these rates will enable. Figures 8.1 and 8.2 show two important graphics from the book. The term HDTV (high-definition television)-based computing refers to the convergence between television and personal computing.

"Professional" applications refer to the extension of corporate LANs into private homes to enable home offices and remote access. "Basic fees" refer to a charge applied to cover a host of applications that might be provided, but for which it might not be possible to charge individually.

Based on this analysis, the key early applications will be corporate LAN and consumer Internet access at high speeds. These begin to be very significant in the 2002 time frame. In the 2005 time frame, video

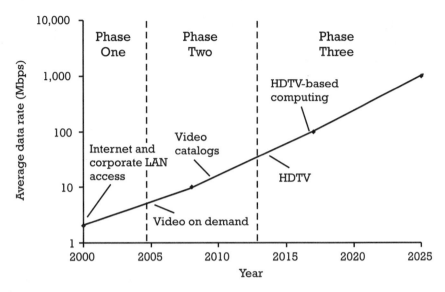

Figure 8.1 Prediction of data rate requirements versus time. (Reproduced from Maxwell, *Residential Broadband*, John Wiley, 1999. Reprinted by permission of John Wiley & Sons, Inc.)

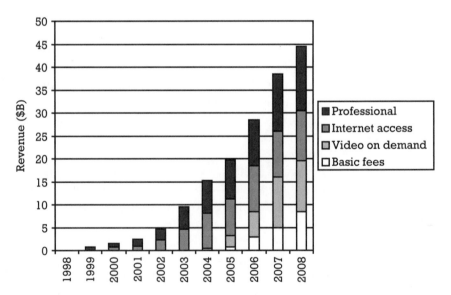

Figure 8.2 Prediction of revenue for different services. (Reproduced from Maxwell, *Residential Broadband*, John Wiley, 1999. Reprinted by permission of John Wiley & Sons, Inc.)

on demand is expected to begin to take off. Video catalogs begin to be significant in the 2010 time frame, and around 2015 HDTV becomes significant.

8.3.3 Wireless consumer applications

For the consumer market, the evolution of wired services to wireless seems to make sense. Wired Internet services have spawned an increasing awareness and use of Internet services in the home and high-speed access in the home is beginning to take off. We are also seeing some Internet access beginning to be important in wireless. NTT DoCoMo's iMode service seems to have finally found the formula for high penetration of consumer wireless services with 11 million users by mid-2000 and 30% penetration of its wireless customer base. Probably most significant is that DoCoMo is beginning to see a significant increase in its average revenue per user from iMode.

A difference between the wired scenario and the wide area wireless scenario, however, is the difference in bit rates. The iMode service is at a significantly lower rate than wired access at home and will continue to lag. Users, however, have shown a willingness to put up with lower performance for wireless voice services than for wired services, and once a threshold is reached they will either choose appropriate applications for wireless or accept the degradation in performance.

Looking at future home applications and services, video on demand appears to be the next major driving application/service in the home. What is not clear is whether providing video to the home will trigger wireless video on demand. There is no doubt that users will want to take this application with them. They will need to accept greater cost and lower utility, however. The arrival of this service is also likely to be some years after that of video on demand to the home.

The next significant application in the home is video catalogs and shopping. Again, this will be another application that people will want to take with them as a form of entertainment or to enable them to shop when they want to or when they have time. Similarly, this application will need to be cost-effective and seamless and will again likely follow shopping in the home. The home space will be the high-volume, high-usage environment that will drive the development of appropriate technology and content. Delivering these same applications to wireless devices when away from the home will follow.

8.3.4 Wireless business applications

The projected business applications for the home seem to make sense as a basis for projecting the evolution of wireless applications and services. Initial cellular users were primarily business users who could justify costs based on improved efficiencies. With the growing trend toward telecommuting and major corporations putting the infrastructures in place to support working at home, these services would seem a natural for evolution to wireless. Certainly, wireless operators recognize business users as a key customer segment that has provided a major part of their system revenues. They recognize the opportunity to serve business users with data applications and services. There have, however, been a significant number of barriers to meeting their needs. Some of these barriers included:

- Numerous data transport standards;

- Applications that were fragmented and not suited to wireless;

- Wireless networks that were too slow;

- Wireless networks that lacked the in-building coverage needed by many customers;

- Wireless networks that lacked the nationwide/international coverage needed by large companies;

- Inadequate security for corporations.

While these problems have not been completely solved, there has been major progress in many of these areas. The Internet with its Web-based applications is providing a common data transport standard. The Web is also providing a common format for corporate applications. Wireless networks are beginning to roll out high-speed packet capabilities. Wireless operators continue to consolidate to provide national and international services, and virtual private networks (VPNs) are beginning to provide adequate levels of security. This, we believe, will lead to a significant early market for data applications and services for business users. Continuing improvements in wireless networks will continue to drive more business applications to wireless.

8.3.5 Other application directions

In addition to being able to provide applications that require a higher bit rate, applications will also become more intelligent. They will be more aware of the user's personal preferences, what the user is trying to do, and the user's current context. Applications will make use of agents to get any help they need in the network to meet the user's needs, and they will provide the results in a way that the user finds natural.

Wireless outside the home can add additional information to make these applications even more capable. Information such as the user's location will provide a much richer environment for intelligent applications. Innovative user interface technologies will be developed to allow natural access to complex information while on the move. Potential solutions include retinal projection and light, flexible displays. Finally, it is expected that there will be significant machine-to-machine communication. This communication will help make various systems more efficient by providing capabilities for sensing performance, problems, and need for maintenance.

Because many of these intelligent capabilities rely on information that is obtained from the local environment, many of these capabilities will evolve over time as wireless sensors, embedded intelligence, and ubiquitous access to a network become commonplace.

8.4 Wide area cellular system considerations

8.4.1 Background

Technology-related limitations will certainly shape the future wireless world. The dominant technology for mobile applications to date has been wide area cellular. As wireless systems evolve to provide high-speed Internet access and other high bit rate services there are key issues that could limit the ability to provide very high speed data. Reviewing some of the key issues for wide area cellular systems as they might evolve to higher bit rates therefore is important.

8.4.2 Spectrum issues

One of the key issues is the availability of spectrum, which can be a limiting factor in the number of customers a system can serve, the cost of the system, and its bit rate. The International Telecommunication Union (ITU) developed a model for spectrum requirements through 2010 based on inputs from a wide community of interested parties including wireless operators and manufacturers. Using this model they projected the need for an additional 160 MHz of spectrum for cellular applications. Plans for the allocation of this spectrum are now in progress. Based on this model, we have generated estimates of the spectrum allocation increase required to support 10X (20 Mbps) and 100X (200 Mbps) peak throughput increases referenced to the current international mobile telecommunications 2000 (IMT-2000) peak data rate (i.e., 2 Mbps). These estimates were not only based on providing higher bit rates but also included improvements in spectral efficiency expected over the next 10–20 years. Based on this analysis, we have concluded that an additional 250 MHz of spectrum will be needed to provide a 10X increase in bit rate, and an additional 5.4 GHz will be needed for a 100X increase. The analysis shows that the additional spectrum allocation requirement is driven by capacity requirements of urban pedestrian users. Details of this analysis can be found in Appendix 8A.

It is also of interest to consider the frequency range for any new spectrum that is needed. The frequency band for new systems has increased from 800/900 MHz for first- and second-generation systems to 2 GHz for third-generation systems, and future systems will be forced to even higher frequencies. It is expected that in the 20-year time frame, wide area cellular systems will move to the range of 3–5 GHz.

Certainly it is obvious that 5.4 GHz cannot be made available for wide area wireless systems in the 3–5-GHz frequency range. This spectrum requirement clearly calls for a closer look at the assumptions and for alternate approaches to meeting needs of high-speed data users.

8.4.3 System economics

Another key issue that will affect the evolution of cellular systems is the cost of the service to the customer. Many factors will influence this, including site identification, acquisition, or rental and construction costs. Other factors include costs of spectrum, infrastructure equipment, subscriber equipment subsidies, maintenance, financing, billing, administration sales, and advertising.

Trying to look at these costs in any detail is beyond our scope, but a closer look at a few of the areas is of interest. First consider wireless equipment trends. Wireless infrastructure and subscriber unit costs are strongly driven by semiconductor trends, which from 1975 to 1995 show a 5,000X improvement in processing power [2]. There is continuing debate on when this trend will slow, but over the past 10 years, despite many predictions of slowing, the trend has held. Current views are that feature sizes can be reduced by another factor of five resulting in a 25X increase in device density and device speed. With these improvements in architectures, it is reasonable to expect that infrastructure and subscriber equipment capable of 100X bit rate will be roughly the same cost as equipment today.

Site costs including identification and construction or rental are likely to be relatively constant on a per-site basis. Systems serving the same number of customers are likely to have relatively similar customer acquisition, advertising, and subscriber equipment subsidy costs. Therefore, one of the major variables in the overall cost of the system is the number of sites required to service a given customer base.

It would be worthwhile to identify the major factors determining the number of cell sites. The ITU report specifies minimum practical cell sizes and uses these to compute spectrum requirements. Our analysis assumed no reduction in cell sizes over the ITU study. We did assume, however, a nearly 10X improvement in spectral efficiency so that little additional spectrum was required to get 10X the bit rate. However, going to 100X, little additional improvement is expected in spectral efficiency, so a large increase in spectrum is required.

Because these large increases in spectrum for 100X bit rate are unlikely, we need to look at other possible solutions. Given that the user density is constant in a certain area, if we reduce the cell area by a factor of 10, the cell capacity can now be shared with one-tenth the number of users. Therefore, we can provide 10X the number of bits to each user without any increase in the spectrum required. With the appropriate system design, this larger number of bits to each user can be delivered at 10X the bit rate. So, in dense urban areas, if we can practically reduce cell sizes, we can trade bit rate increases directly for number of cells without increases in the amount of spectrum required.

Moving away from the dense urban areas, cell size can be increased because the number of users in a given area is reduced. At some point, we will no longer be able to increase cell size because it will be limited by the range of the site. In these areas, the number of sites is determined largely by the range of the system. Range is determined by the bit rate and the propagation characteristics at the frequency used by the system. Table 8.1 shows the results of our analysis on this topic.

As with any analysis of this type, there are many assumptions. The system design was based on a high bit rate packet data design where a channel is shared by a number of users through a time domain access scheme on both inbound and outbound paths. For the same data rates on both inbound and outbound paths, this approach would mean that the cell spacing would be determined by the inbound path. This design, however, is based solely on the outbound path, and the inbound path

Table 8.1
Cells Required for 10-Kbps, 800-MHz Systems Versus End-User Bit Rate

Bit Rate (Mbps)	800-MHz Band	2-GHz Band	4-GHz Band
2	1.1X	3X	6X
20	3X	8X	16X
200	10X	30X	60X

supports a reduced rate to achieve the same range. In addition, the system design anticipates a 10–12 dB improvement in system gain over the next 20 years through various means such as adaptive antennas. These assumptions reduce the number of cell sites dramatically for the high bit rate system, but, as seen in the table, the increases in number of cells are still very large. As discussed in Section 4.3, small cells bring their own problems, in particular the high cost of backhaul from the cell.

8.4.4 Conclusions

It appears that a significant increase in the number of cells is required in high population density areas to limit the need for large increases in spectrum. In low population density areas, the number of cells also needs to be higher than it is today, since providing a higher bit rate requires a move to a higher frequency band. This points to the following types of solutions:

1. In urban areas, deploy a technology that can support small cells and that can have low cost and small base stations that are easily deployed.

2. In rural areas, maximize the cell size by using lower frequencies and provide lower bit rates.

A bit rate of 20 Mbps to an end user appears to be viable in urban areas without increasing the number of cells and with only a moderate amount of additional spectrum. Therefore, it would appear that the cost per bit to the end user will be reduced by an order of magnitude as these systems become widely deployed (2007–2010). This should stimulate the use of new services. Going to 200 Mbps, hovever, will require on the order of 10X the number of cells in dense urban areas. This site density increase will likely result in either increased cost/bit or a significantly different type of system.

8.5 Nomadic and fixed networks

Two alternative technologies are being actively developed to extend the reach of wireless systems: nomadic [i.e., W-LANs and wireless personal area networks (W-PANs)] and fixed. These systems are designed to deliver broadband services at higher data rates than is possible using wide area mobile technology. This capability is enabled by dispensing with key requirements of mobile systems such as seamless handoff, wide area coverage, and mobile operation. The time to market and throughput advantages of nomadic systems are shown in Figure 8.3.

Figure 8.3 depicts the time frame for initial deployments of W-LAN and wide area mobile technologies. Note that W-LAN systems have approximately a four-year time to market advantage at a given data rate and a 25–50X throughput advantage at a given point in time. These advantages may well prove to be decisive as we move into the broadband future world. There are, however, significant hurdles that W-LANs must

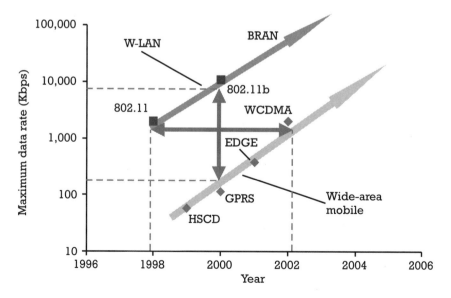

Figure 8.3 Wireless systems comparison.

overcome to become a core part of the wireless broadband future. These issues are discussed in the following sections.

8.5.1 Nomadic systems

A nomadic system provides wireless coverage only in focused coverage areas. Thus, these systems do not attempt to provide wide area coverage. Within the coverage area a nomadic system will allow for modest movement of the terminal in terms of distance and speed. By this definition, nomadic includes both W-LAN and W-PAN type systems.

8.5.1.1 Wireless local area networks

W-LAN technology is being developed to meet emerging needs for untethered access to home and business networks. Two families of standards dominate work in this area, 802.11 in the United States and HiperLAN in Europe. The 802.11 systems have been standardized with maximum throughput rates of 1–11 Mbps at 2.4 GHz and up to 50 Mbps in the 5-GHz band. The HiperLAN family is being designed to provide up to 54 Mbps throughput in the 5-GHz band. These systems are designed to operate over much shorter terminal-to-site distances than for wide area systems. Ranges of 100–200m are the norm.

W-LAN systems currently have far less mature standards than do wide area systems. Vendors, however, are realizing the indispensability of interoperability and volume pricing in their product offerings. Therefore, we expect nomadic system standards to improve over time.

Although W-LAN system sales are expected to grow over time, by mid-2000 their deployment was limited to a few applications such as factory automation. One of the key applications spaces for wireless LANs in the future is expected to be the home. As video on demand services spread into the home, there will be a desire to simplify installation. Typical homes in the United States have multiple TV sets throughout the home, and with older houses the installation costs can be significant. Operators, however, will have the option to use W-LANs for distribution. As W-LAN technology becomes less expensive and easier to deploy, it will be more cost-effective to install a W-LAN in the home rather than wire the home. This application is expected to drive W-LAN technology into the mass market. As W-LAN technology begins to penetrate the home market, other types of devices will be introduced with this capability; computers and related equipment, video cameras, still cameras, audio equipment, cell phones, and other devices will all provide this capability to allow easy interconnection in the home. Once low-cost W-LANs are available and many types of devices have this capability, the opportunity will exist for these W-LANs to move out of the home.

8.5.1.2 Wireless personal area networks

PANs are an emerging technology that allows the wireless interconnection of devices over short ranges. The ability to provide wireless connection for devices is a key enabler for many important applications.

The first implementation of these types of networks, called Blue-Tooth (IEEE 802.15), allows bit rates near 1 Mbps at ranges up to 30 feet. It supports both point-to-point and point-to-multipoint applications. Some of the initial applications are expected to be wireless headsets for cellular radios, thus separating the cellular phone's human interface from the wireless modem.

A number of additional applications are envisaged. Another application for a cell phone with an embedded PAN device is as a cordless phone. This device could use PAN communication to an in-home access point and transfer to cellular when outside the home. It is also envisaged as a technology that can be embedded in a computer. The computer can then talk to a user's cell phone or another computer or network without any wires. The technology also allows a number of users to form a small network to share information within the group. There are many more applications that this technology will enable and it is expected that this technology will become ubiquitous.

This ubiquity will be driven not only by the fact that there is a standard but also by the expected low cost—below $5 by 2005. This technology has generated a lot of excitement in the industry, and by 2005 virtually all cellular phones are expected to integrate it into their handsets. This integration provides the PAN device with a user interface, power source, and application platform. The huge volumes and customer base of cellular phones is expected to further drive the costs down.

The other major impact of its integration into cellular phones is that now the technology becomes mobile. Because a large percentage of people will have the capability, the technology can be implemented into many other devices. It can be used to put money on phones (although security will be a key issue). The phone can then be used to buy things at stores or from vending machines through the cell phone's browser. It can be used to access information available at public access points. One possibility would be to access detailed product information at an access point in the store. It can also be used to exchange information in a public setting for such applications as expediting airport check-in. Within 10 years, we would expect that it will be an indispensable part of our lives.

Over time, it is expected that this technology will migrate to higher bit rates. Digital still and video cameras and computers will need higher than the 0.7 Mbps initially available. In mid-2000 the 802.15 committee started to standardize a 20-Mbps data rate. Future home entertainment devices (such as HDTV) will drive up the rates even farther in the future to greater than 100 Mbps.

This technology will be embedded in many devices and will be used in many different environments to provide a simple, easy to use connection to other devices and networks.

8.5.1.3 W-PAN/LAN integration

It would seem that there is a strong overlap between W-LAN and PAN capabilities and applications. This overlap will potentially favor the more rapid development of the technology that is deployed first. Devices with PAN capability are expected to be available in the very near future and some of the initial products will be cell phones.

W-LAN capability in consumer devices is expected to lag the introduction of PAN devices because of its significantly higher cost. This earlier introduction would seem to give the advantage to the PAN devices. PAN devices, however, do not currently have sufficient range or bit rate for many future home applications nor the full networking capabilities needed. This need for continuing evolution of standards provides the opportunity to combine the capabilities of W-LAN and PAN devices into an integrated standard, which could allow sharing of much of the hardware and software, thus minimizing the costs. Given the continued trends predicted by Moore's Law and the driving forces in the home, W-LAN capabilities can be expected to proliferate. PAN costs will be a small increment over W-LAN. Therefore, we would expect the PAN and W-LAN capabilities to be integrated over time. In Europe, standards are already under development for an integrated solution via HiperLAN and HiperPAN.

Over time, these capabilities will be integrated into wide area wireless devices. This development will allow customers to access whichever system is most convenient or suited to delivery of the desired application.

8.5.2 Fixed systems

Fixed terrestrial broadband systems are also emerging. These systems can compete directly with wireline solutions such as xDSL and cable modems or complement them by serving customers difficult to access by wireline. The current lack of maturity in this market space, however, has resulted in a chaotic environment, with solutions using numerous technologies, deployment topologies, and frequency bands competing for dominance.

Frequency bands for fixed systems around the world include 2.5–2.7 GHz (mostly in the United States and Latin America), 3.4–3.6 GHz (widespread outside of the United States, especially in Europe), 5.4–5.7 GHz (mostly in the United States), 26–28 GHz (mixed, with most countries allocating part of this band), and 38–43 GHz (also mixed with most countries having, or considering, allocations somewhere in the band). Possible modulation technologies included constant envelope (i.e., FM), single carrier QAM, and OFDM QAM. Finally, two system deployment topologies are being considered, point-to-multipoint and mesh. Point-to-multipoint is the familiar cellular concept with a site at the center of a cell serving a large number of subscriber units. A mesh topology dispenses with the site entirely. Rather, connectivity is attained between subscriber units and a network access point via ad hoc networking. Thus, traffic is wirelessly routed by hopping from one subscriber to another until the access point is reached.

Standards addressing this space are under development in the United States (802.16), Europe (BRAN), and Japan (ARIB). In 802.16 and BRAN, separate standards development is under way for above and below 10-GHz systems.

This class of wireless access technology has the potential to capture a significant portion of the home and business "last mile" data market. However, significant obstacles must be overcome, including: significant lowering of equipment cost, improved system availability, improved time to market, and reduced installation costs.

We believe that fixed systems will play a major role in "last mile" data delivery to homes and businesses in the mid-term (i.e., 2010–2015). Longer term, the penetration of fiber and high-speed wired solutions might well dominate the market. Fixed wireless solutions will continue to address those areas difficult to reach with fiber and cable.

8.5.3 Wireless broadcast technologies

Broadcast technologies have a long history in both radio and TV. Of particular interest is TV. Until recently, TV broadcasting has been based on analog technologies and has provided relatively few channels. Digital TV with capability for significantly more TV channels has been defined, but so far there are few digital TV broadcasters. In the meantime, cable

TV has made steady inroads into the home market. This has left the TV broadcasters with a declining market. While they may eventually try to recapture the market, it will be difficult to recapture their viewer base. This is especially true since the broadcasters did not have a strong customer relationship. In addition, competition is heating up with xDSL from the phone and cable companies, which will be able to provide two-way services such as telephone, Internet access, and other video services on demand. While broadcast will not go away, it will probably continue to be relegated to a secondary future role in the United States.

The future seems a little better for broadcasters in Europe. They appear to be ahead in the race for the home market. They are in the early stages of providing access to many digital channels and are selling these services to end users. This should provide them with a strong customer base, which they can exploit. Should they continue to lead in the race, they are likely to be a force in the home market. There has been a lot of discussion about how they might converge with wireless systems. One option is to have a hybrid system that uses broadcast to a subscriber device and cellular for the return path. Since broadcast is a very efficient means of distributing common information to a large group of users, this could be viable in the near future. We believe that over the long term, the desire to get customized services when required and the increased availability of wireless transport capacity will relegate broadcast to a secondary role. If the European broadcasters get an early head start in the home market, however, they could exploit their customer base to help them move into two-way services and still be a significant force in the evolving wireless market.

8.6 Business, regulatory, and standards

8.6.1 Regulatory and standards

One of the key regulatory driving forces is the World Trade Organization's Telecom Agreement [3] that took effect in January 1998. The agreement was initially signed by 55 countries that represent over 90% of the global telecommunications service revenues. This agreement allows direct and indirect foreign ownership of telecommunications infrastructure, provides competitive safeguards, reduces the power of dominant telecommunications operators through interconnection, regulates licensing criteria, and provides for international enforcement. It covers voice telephony, data transmission, fiber, radio, and satellite. Clearly, the intent is to open the global telecommunications market to international competition. We are already seeing an increase in foreign investment and partnering by major telecommunications operators, as well as significant global operator consolidation, and the trend seems to be accelerating.

8.6.2 Telecommunications operators

Over the years there has been a continued consolidation of wireless operators in the United States as the market has matured. With growth, wireless operators started to reach the consumer end of the market, forcing a continuing focus on reducing costs while maintaining the average revenue per subscriber. This resulted in a continuing push for consolidation and it has also driven changes in the approach to pricing wireless services. Rate plans shifted from relatively constant cost/minute plans to packages with significantly larger numbers of minutes at a much lower cost/minute. This approach reduced the cost/minute but allowed the operator to maintain or increase revenue/subscriber. By 2000, a significant shift in minutes of use from the wired network to the wireless network was occurring as a result of these rate plans, and this is likely to continue over a number of years.

Wireless operators, however, cannot be regarded in isolation. In many cases, at least in the United States, they are owned by local telephone companies. In addition, most major long-distance companies in the United States have also acquired wireless properties. For instance, AT&T owns cable, long-distance, and wireless operations.[1] South-Western Bell Corporation (SBC) owns cable, local access, and wireless properties. Internationally, the major telecommunications operators own both local access and wireless in many cases. The large telecommunications operators are continuing to utilize their strengths to buy properties in other countries. As these telecommunications operators expand into the home and offer entertainment, Internet access, and wired phone services, they can gain advantage by bundling. This diversification into other access technologies allows operators to increase their customer base and reduce their costs. Owning other access methods also reduces the risks that new operators who participate in only one of the access methods will be able to attack the diversified telecommunications operator.

---

1. However, at the time of writing in March 2001, it was considering breaking up into a number of discrete operating companies.

A continuing driver for the telecommunications operator will be the cost of infrastructure. As discussed in Section 8.4, higher bit rates will require significantly more wireless infrastructure. In addition, operators that are moving into other access businesses (cable, DSL, fixed wireless) will have major capital investment requirements for these businesses as well. The advantages of providing integrated access should be significant to operators. Not all operators, however, will be able to afford the capital expenditures. In addition, the more operators in a given area, the more fragmented the area and the more expensive it is to serve. Over time these factors are expected to result in continuing operator consolidation. As other markets have matured, the result has typically been a maximum of three major players per market. The telecommunications operators would be expected to undergo a similar transition over the next 20 years. The telecommunications market will also become an international market as driven by the World Trade Organization Telecom Agreement. Another key issue for telecommunications operators is the need to differentiate themselves from their competition. For wireless, data services have been considered as a way of achieving this. But as noted earlier, there have been a number of barriers to the popularity of these services. With the emergence of iMode, there is a stampede to try to take advantage of the potentials of this new service.

Taking advantage of this new service, however, requires a significant shift in the operator business model and their approach to their business. Operators to date have had a highly vertically integrated business controlling all aspects of the service. They own the spectrum, the sites, and the equipment. They provide sales, marketing, distribution, billing, and maintenance, and they provide the service (voice communication). Through this model they own the customer relationship. The Internet will cause a major change in this model. Instead of voice communication being the only application, there will be many services provided by a wide range of content and service providers who will utilize their strengths and customer relationships in the wired Internet world to move into wireless. As other services such as video on demand become popular in the home, these service providers will use their customer relationships and provide these services on wireless. These Internet and other home content providers are therefore expected to gain more power with wireless operators as their services become more popular and they are viewed as a key element of the wireless service.

Over the long term, as end users want more control of their own applications, services, and personal information, this could lead to new players that might provide personalization services independent of operators or content providers. In any case, in the face of intensifying competition, it seems unlikely that telecommunications operators will dominate end-user services.

In 20 years, then, we would expect that there will be fewer telecommunications operators and that they will play in a number of different access technologies. Driven by the World Trade Organization's telecommunications agreement and local regulations, telecommunications operators will be required to provide generic transport services to a broad range of end-user service providers. However, they will also build on their strengths to partner with content providers and others to maintain some differentiation with their own branded value-added services.

8.7 Public acceptance

Over the past 20 years there has been an unprecedented acceptance of technology in general, and wireless in particular, as the vehicle of progress. Thus, were we to simply extrapolate from the past, the future would look rosy. Even in the midst of this enthusiastic acceptance, however, undercurrents of suspicion and discontent are discernable. Voices have been raised questioning the health, aesthetic, safety, and privacy implications of wireless communication, among others. Clearly, the current majority presumption is in favor of wireless communications. The challenge for the wireless industry is to continue delivering high value services while also taking seriously the issues that threaten its viability.

It is natural for those of us with scientific and engineering training to simply discount the suspicion and hostility of some toward wireless technology. Indeed, in a large proportion of cases these concerns are clearly based on false assumptions and information. We must, however, keep in mind a couple of points in order to maintain a proper perspective, those being:

1. Just because critics are often wrong doesn't mean that they are always wrong.

2. The underlying values of these critics speak to human concerns that are shared by many who still choose to use wireless technology.

Therefore, we would do well to take inventory of these perceived risk factors.

8.7.1 Health and safety

Health and safety concerns surrounding wireless devices map almost exclusively into fear of cancer and traffic accidents, respectively. A secondary safety issue is cell phone use on airplanes.

8.7.1.1 Current health environment

Most news stories on cellular health effects provided clear statements from key governmental and/or industry groups (i.e., the FDA, WHO, and CTIA) to the effect that "there is no current evidence of a public health threat from wireless phones or towers." However, they also tend to give equal weight to empirically unsubstantiated claims or concerns about cancer risk from members of the public. A specific example from the *Boston Herald* [4]: "My girlfriend is a biologist and she refuses to call me because she's afraid she's going to get brain cancer."

The wireless industry has done an excellent and responsible job of funding independent research on this issue and communicating results to the public. More recently, the industry has gone to the next level of disclosure by publishing radiation absorption data on their cell phones. This strategy of empirical science and full disclosure is sound as long as the public at large can be characterized as follows:

1. Has confidence in the independence of scientific studies;

2. Has confidence in the ability of science to address these issues;

3. Is convinced of the value of wireless products and services.

Of course, we are implicitly assuming that good science will continue to overwhelmingly support the safety of wireless products. Another prudent industry strategy would be to fund R&D on minimizing energy absorption by the user and moving the resulting technology into the product.

Finally, it must be noted that the emissions from cellular antenna towers are also regularly cited as posing health risks, although less often and with less intensity than for handsets. These claims often accompany complaints about the aesthetic and property value impact of towers on a neighborhood.

8.7.1.2 Current safety environment

There appears to be near universal agreement in the media that wireless phones pose a major safety risk on the road. Cell phone users appear to be singled out as particularly dangerous even though this is just one of innumerable sources of distraction for a driver.

In these media stories the wireless industry rarely, if ever, appears to defend itself. Occasionally, historical incidents of distraction concerns that were eventually proved overblown are cited (i.e., windshield wipers and car radios). There is also the occasional acknowledgement of other forms of distraction.

This lack of industry comment is understandable in that we have all seen examples of dangerous driving while using wireless products. We cannot and should not ever defend the unsafe use of our products. It is also clear that using a phone, even hands-free, will lead to some distraction of the driver, which may reduce the safety of their driving. The best long-term solution will be to work closely with the automobile industry to create less distracting user interfaces. Work to detect and prevent driver overload may also yield beneficial results.

8.7.1.3 The future

The key question for 20 years out is whether the public at large will become more or less sophisticated in its ability to evaluate the complex scientific information underpinning wireless communication services as it relates to health and safety. If the answer is less, then the industry could find itself under sustained attack in the not too distant future. The more likely scenario, however, is that good science will continue to drive out false information. The key challenge for the wireless industry is to continue the funding of independent, high-quality and wide-ranging research on these issues. The already massive body of scientific data clearly shows wireless products to be biologically safe, and it should be possible to make intelligent use of communications in the car of limited distraction. We simply must stay ahead of the curve in terms of being ready to answer new concerns with credible information.

8.7.2 Aesthetic

8.7.2.1 Current environment

Aesthetic concerns coalesce primarily around the placement of wireless sites in neighborhoods. The main issue appears to be that of lowered property value. In most cases, the issue of health risk is raised along with the aesthetic impact.

Operators in these disputes tend to stress that many sites can be made unobtrusive via camouflage. In cases where overt towers are necessary, however, the dispute often becomes an open power struggle.

The current dispute over cellular towers occurs within a larger societal context. Members of the public are becoming more aggressive about the enforcement of their aesthetic preferences. In the words of a July 13, 2000, *New York Times* article [5], "... more and more Americans are demanding a world free of 'visual pollution.' Appearance is getting the sort of regulatory scrutiny once reserved for public health and safety." And why not?

8.7.2.2 The future
The wireless industry will be buffeted by the growing radicalism surrounding aesthetic issues. Traditional towers may have to be abandoned entirely in favor of virtually invisible micro and pico sites. The economic impact of these changes, especially in rural areas where large cell radii are important, could be very negative unless extremely cost-effective radio, enclosure, power, backhaul, and network solutions are developed.

8.7.3 Privacy
Wireless users are becoming increasingly aware of the potential for loss of control and privacy in their use of these services. The growth of wireless e-commerce will surely sharpen these concerns.

8.7.3.1 Current privacy environment
The ability of government law enforcement agencies to eavesdrop on our wireless communications appears to be the current top privacy concern. A recent federal appeals court ruling has supported the concerns of privacy advocates against the encroachments of Big Brother.

In Europe the privacy issue revolves around the right of individuals to own their personal data. Thus, operator ownership of personal profiles is the primary concern.

Another emerging major concern area is the impact of wireless location technologies. Wireless location service vendors such as SignalSoft Corp. are working hard to convince the public that effective privacy protections are being built into their solutions.

Finally, it is regularly commented on that cell phone users have an annoying tendency to "yell out" private personal information in public places. This would appear to be simply another of the many forms of exhibitionist behavior that appears to be on the rise in many cultures.

8.7.3.2 The future

The demand for privacy is defended by highly organized, effective organizations. There is also clearly significant public support for the maintenance of reasonable levels of privacy. Therefore, it is difficult to imagine the implementation of public or private systems that are capable of violating personal privacy apart from criminal investigations. Thus, although the wireless industry will have to assure acceptable privacy to its customers, it is unlikely to face external threats from government, at least in the first world.

8.7.4 Community

A fundamental concern about communications technology is its impact on human community. The particular issues of concern are isolation, lessening of consideration for others (i.e., etiquette), education, and family, among others.

A general theme that arises from current articles on wireless users is that they are insensitive, arrogant, and annoying. More than likely, a small minority of particularly egregious offenders is responsible for this problem. Unless generally accepted wireless service etiquette can be created, however, this problem is likely to grow worse over time.

8.7.4.1 Education

One of the advantages ascribed to the digital revolution is mass customization. Will this same trend hit education? How will this affect primary education? On one hand, it may allow "units of one" to be a reality for children, having educational programs specifically designed for them and delivered any time anywhere. On the other hand, this individualization may retard social development.

8.7.4.2 Cultural divide

The increasing availability of the new technology may exacerbate cultural divides as easily as it heals them. Certainly to this point the increasing presence of technology has happened as the gulf between the world's richest and poorest has widened—while causality is not clear, it is possible that these economic divides between haves and have-nots will greatly affect the world in the next several decades [6]. Will technology help or hurt? Other differences may emerge as important as well—cultural, gender, age, race, and so forth.

8.7.4.3 Isolation

With increasing communications technology capability comes the potential for people to communicate without being physically present. This brings up the impact of increasing physical isolation on human relationships. Will we actually be more isolated? Early indications are that people who spend more time on-line spend less time in actual face-to-face communication. While much has been written about on-line communities, evidence exists to suggest that these communities are not a substitution for physical communities. On the other hand, will increasing technology capability increase quality of distance "presence"? It isn't yet clear which way this will go.

8.7.4.4 Family

As the ubiquity of electronic communications increases, how will we balance the increased accessibility with the desire to protect our children and provide age-appropriate stimulation?

Pundits are heralding the dawn of a new age of children interacting with other children and adults over the Internet, learning in new ways. In particular, Tapscott in [7] asserts that the "N-gen" will become very adept at discerning reality from fantasy and become expert "editors" of information at much earlier stages than their parents. This is not consistent, however, with the literature on child development—children up to a certain age do not possess the mental capacity to accomplish some of these tasks. Thus, parental safeguards and regulation are probably more important in the digital future, rather than less.

8.7.4.5 The future

There is significant tension between the positive and negative social impacts of wireless communication technology. Currently, the predominant conclusion is that the positives significantly outweigh the negatives. If the general public begins to believe that these services are making society less civil, secure, and livable, a significant backlash could very easily occur.

The wireless industry would do well to institute a coordinated, sustained campaign to encourage community affirming and enhancing use of their services. Another positive step would be to integrate human factors studies into the design of all products and services.

8.8 Future scenarios

This section explores potential future scenarios for the wireless market. By so doing we can test the plausibility of potential futures without claiming the (implausible) talent of fortune-telling.

We believe that end users will demand the creation of unified, ubiquitous wireless communication systems that deliver data-rich content on demand with low latency and high throughput. Typical untethered data rates of 10–25 Mbps will be expected. In the home and business, typical rates of 200 Mbps will be the norm. In addition, they will demand that the number, form factors, and user interfaces of these devices be minimized to the point that utilization is radically simplified.

Given the market demands, application requirements, usage trends, and system economics, the question as to which solution will dominate the wireless space 20 years out remains to be addressed. In doing so, we must accept that we are moving beyond the quantifiable and into the speculative mode of thought. This is the case because only what exists can be measured. What might be must be evaluated via alternate modes.

One particularly valuable means of exploring what might be is through the use of scenarios. We define the term "scenario" as a specific, plausible sequence of events that instantiates an evolutionary path to reach a potential future.

Thus, scenarios allow us to consider the existence of numerous potential futures and to explore the paths by which these futures might be brought into actual existence.

We will posit three scenarios in this discussion:

1. Mobile dominates;

2. Nomadic dominates, where nomadic is the use of wireless in selected, often urban, environments, typically when stationary;

3. Integration dominates.

Note that these three scenarios relate to the type of wireless *equipment and networks* that come to dominate the market. It is completely possible that today's wide area mobile operators could evolve their networks to nomadic dominant. Of course, it is also possible that new operators will enter the nomadic space and rise to challenge the dominant position of wide area mobile operators. Figure 8.4 provides the context for the discussion of potential scenarios.

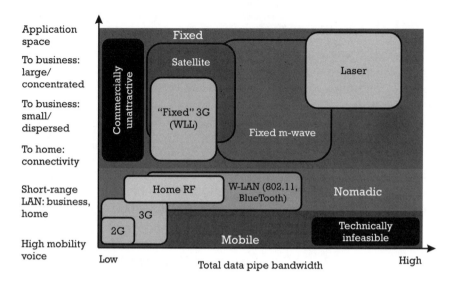

Figure 8.4 View of the wireless world.

Note that in Figure 8.4 the current total wireless world has been partitioned into three distinct "spaces": mobile, nomadic (includes W-LAN and W-PAN), and fixed. We will develop our scenarios using this worldview as our starting point.

8.8.1 Mobile dominates

8.8.1.1 Scenario description

In this scenario, wide area, high mobility solutions continue to dominate and effectively limit the deployment of W-LANs. This is accomplished by the development of new mobile systems that provide the throughput and latency performance demanded by end users. The system is capable of being economically deployed within buildings and to homes and businesses. This feat is accomplished while using essentially the same air interface for all applications. Subscriber unit form factors are successfully developed to meet a wide range of applications, including personal computing, Internet access, and voice and data communication. Figure 8.5 depicts this scenario.

Wireless PAN capabilities are fully integrated into all handsets. This capability provides for all of the significant local communication interactions.

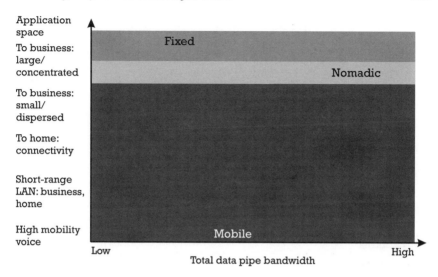

Figure 8.5 Mobile dominates.

In this potential future, customers conclude that the full mobility wireless device is the best portal for the vast majority of their needs. Device capabilities evolve to enable a wide range of services from simple audio to broadband data and video. Wireless minutes of use (MoU) growth continues and accelerates at the expense of wireline, thus providing operators with the market muscle to ward off encroachments by nomadic systems.

The end result 20 years out is a communication culture built around the concept of total mobility and availability. New generations of mobile systems with ever greater capabilities arrive at regular intervals. W-LAN solutions remain in their current niche vertical market segment or disappear entirely.

8.8.1.2 Scenario analysis

For this potential future to become a reality, a number of formidable obstacles must be overcome. First, mobile systems must be capable of serving in-building markets at costs competitive with W-LANs. Second, mobile systems must be able to cost-effectively (e.g., without an explosion in the number of cell sites, particularly outdoors) move to higher per user throughputs, potentially up to 200 Mbps. Third, large new spectrum allocations to mobile systems may be required. As has previously been pointed out, these obstacles are likely to be difficult to overcome.

Given the very large and growing penetration of mobile devices, this solution starts the race with significant advantages. This base can be leveraged in the spectrum, standards, equipment, and marketing arenas to powerful effect. Based on our analysis, however, it is very difficult to envision a plausible path for wide area mobile systems to deliver the wireless broadband solution 20 years out.

8.8.2 Nomadic dominates

8.8.2.1 Scenario description

In this scenario new classes of subscriber devices based on W-LANs reach a level of penetration that is similar to wide area devices. The growth of W-LANs is initially driven by the need for simple installation and mobility of high bit rate services in the home (video on demand). As more W-LANs are installed, many devices begin to provide W-LAN capability. These include video cameras, digital still cameras, audio equipment, computers, printers, and other related equipment and wide area wireless devices.

An operator without a strong wireless portfolio steps in to provide W-LAN access points to business offices, malls, stadiums, and other public areas, leveraging the large base of W-LAN capable devices. These deployments grow to create a system so widely deployed that it begins to appear to be ubiquitous even though coverage is not really "wide area" as we currently understand this term. The applications that dominate usage growth end up being better suited for nomadic delivery and thus, nomadic systems begin to displace wide area systems in terms of high value usage. The much lower costs and higher bit rates possible with W-LAN technology compared to wide area wireless also provides a strong driving force for increased nomadic use of services available in the home. Thus, wide area systems become adjuncts to the more highly utilized nomadic solutions, primarily providing voice communications.

Residential applications that drive large market penetration include entertainment (video, music, games), home information (stove, washer/dryer, security, schedules), family communication (intercom, telephone), and remote computing (e-mail, Web browsing, spreadsheets). The remote computing capability also creates the demand for convenient network access in business situations. Thus, office buildings, conference rooms, airports, hotels, and other public spaces with significant concentrations of business professionals install standardized wireless LAN equipment. Children and young adults begin to expect the wireless delivery of games in their homes, neighborhoods, and popular gathering places. A whole new class of games arises that exploits the untethered nature of wireless gaming. Wireless LANs are also utilized to provide value-added services at sporting events. Finally, malls and stores provide value-added services to enrich and simplify the shopping experience.

This transition may be jump-started by the prevalence of BlueTooth-based devices and services within the next few years. It is currently estimated that by 2004 around 1 billion BlueTooth chip sets will be sold. This rapid growth and accompanying ubiquity of untethered services has the potential to completely transform the nature of wireless communication.

These developments drive the creation of whole new classes of terminal devices. General-purpose devices suited to untethered entertainment and information delivery become commonplace. Laptop computers are shipped with W-LAN capability as a standard feature. Wide area wireless handsets also are shipped with suitable W-LAN capabilities.

The end result of this revolution 20 years out is that nomadic equipment dominates the wireless market in terms of data throughput and revenue generation. Thus, nomadic solutions push into much of the existing fixed and mobile market space (see Figure 8.6). While mobile and fixed systems continue to exist, they are relegated to secondary roles in the overall wireless market.

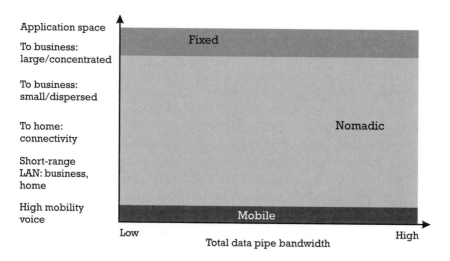

Figure 8.6 Nomadic dominates.

8.8.2.2 Scenario analysis

The above scenario becomes plausible only if the following conditions are met:

- W-LAN technology becomes widely deployed in the home.

- W-LAN and W-PAN systems become strongly standardized with only a couple of dominant standards in each class.

- Consumers place a high value on untethered delivery of a wide range of services.

- W-LAN systems can be designed and economically deployed that have the capacity to service high-bandwidth and low-latency applications such as entertainment and games.

- Intersystem handoff between mobile and nomadic systems is enabled.

- Market economics allows for the build-out of nomadic systems in a wide variety of venues.

- Intercell and system handoff is enabled for nomadic systems.

- Mobile systems do not capture the in-home/building space first.

- New, successful classes of applications emerge that exploit untethered usage and that are uniquely enabled by nomadic systems.

Nomadic systems do have some significant advantages. Most importantly, by dispensing with the requirement to support high-speed movement and wide area coverage, they have attained a significant lead in throughput. This lead is likely to be sustained into the future. The cost structure of deploying nomadic systems in buildings may also be very favorable as compared to wide area systems. Finally, the existence of a large and valuable incumbent market may slow the evolution of mobile systems toward more general information delivery services.

Because of the massive resources and device penetration of current mobile systems, they clearly have a major initial advantage. If wide area mobile systems can be efficiently adapted to meet the new information society requirements, it will be very hard for nomadic solutions to break this stranglehold. However, as we have previously concluded, mobile systems will be unlikely to pull this off.

8.8.2.3 Alternative scenario

The above scenario for W-LAN domination has a telecommunications operator playing a major role in driving the scenario. This scenario could also evolve more in line with the evolution of the Internet. In this scenario, as the telecommunications operators continue their consolidation, regulators move to make sure all potential service providers have access to basic transport facilities in accordance with the World Trade Organizations Telecom Agreement. During this process, standards and regulations are introduced that ensure fair access to a common set of basic transport services at a regulated price based on costs. As part of this agreement, approaches to network configuration and management are defined that allow the simple addition of new access nodes. Approaches are also defined that allow an access point owner to sell access back to the network operator based on its usage. These regulations and the continuing reduction of costs for W-LAN and BlueTooth access points cause many people and small companies to install their own access points in their homes and public places. They reduce their own network charges by allowing public access through their access point. The growth in access points quickly becomes exponential and coverage becomes nearly ubiquitous.

An impact of such a scenario is that it encourages the addition of sensors and many other types of devices to the network. Initially, this allows the addition of sensors by companies that want to monitor the environment to control and optimize business processes. It quickly spreads to the home to allow remote monitoring and access. This culminates in the sensors being placed throughout the public environment with many of the outputs and connections freely available to the public. These form a ubiquitous sensor and intelligent infrastructure that provides not only network connections but also a broad source of environmental information and embedded intelligence that can be utilized by end-user applications.

8.8.2.4 Alternate scenario analysis

The above scenario becomes plausible only if in addition to the conditions noted in the last scenario, the following conditions are met:

▸ Regulators move strongly to drive the broad availability of access to a set of common network services at a low cost, allow individuals to supply access points for the network, and allow access point owners to sell access back to the network owner.

▸ Simple methods are developed that allow easy installation, network management, and microbilling for access nodes.

This scenario is a very interesting one and has potential to generate a significant discontinuity in the telecommunications businesses. In order for it to occur, there are some significant regulatory, technology, and other barriers to be overcome. Considering, however, what has happened in the evolution of the Internet, it cannot be dismissed.

8.8.3 Integration dominates

8.8.3.1 Scenario description

In this scenario, wide area wireless standards are integrated with nomadic standards to provide for a single integrated capability. A user can seamlessly access applications and services via nomadic and mobile networks using a single device. The system performance will vary with the system and environment. The "look and feel" of the services, however, will remain the same. This situation is depicted in Figure 8.7.

One way this future could evolve is through the continued merger of wireless and wired operators. In this scenario, wireless operators continue their consolidation into a few powerful companies. These operators have evolved from other telecommuncations businesses, and they realize the importance of wireless and the emerging market for high bit rate services to the home. They typically own a backbone network—wireless

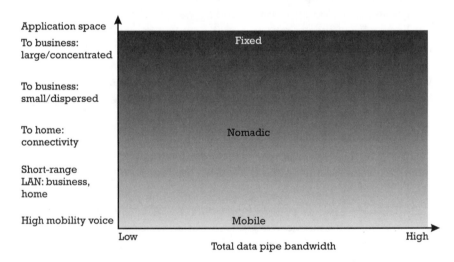

Figure 8.7   Integration dominates.

and cable or local telephony (migrating to DSL). As they participate in the home market, they see the need and the value of W-LAN capability in simplifying installation. They also see the drive for higher bit rates to the home to provide for new classes of applications. In the wireless world, they see the need for higher speed driven initially by Internet access. They also see the potential to provide the home services to their customers when they are outside. However, understanding the limitations of wireless in providing very high bit rate services, they decide to drive standards development that can allow W-LAN technology to be highly integrated with wide area wireless technology. Their significant presence in both wide area wireless and high bit rate home access provides a powerful force in the standardization process. Each type of system is deployed where appropriate. This allows operators to build upon their home customer base to provide a new source of revenue while offsetting some of the revenue lost when their wide area wireless customers use the lower cost W-LAN infrastructure.

Continued loyalty depends on operators providing the most seamless delivery of the broadest range of services. The economics of consolidation enables operators to maintain profitability in a world of highly distributed application and service provision. This is accomplished by the development of applications and services that are branded by the operator and optimized for their network. Thus, customers look to the operator first as a source of services as well as connectivity. Finally, the sheer volume of traffic generated by highly valued services allows the "data pipe" portion of the business to remain reasonably profitable.

The implication of this potential future is that nomadic system types will emerge as important components of overall wireless access. Thus, we would expect that the growth in sales for these system types would be greater than for mobile systems over the next 20 years. Mobile systems may well maintain sales dominance in absolute terms because of their huge existing market base. However, it is not inconceivable that, due to changing use profiles, nomadic equipment could capture a very substantial portion of wireless equipment sales. This occurrence would require the broad rollout of these systems in homes, businesses, hotels, airports, and shopping malls.

8.8.3.2 Scenario analysis

Looking at the current situation, we can already see instances of operators providing wireless and high bit rate services to the home. Certainly AT&T is a prime example with its purchase of TCI. SBC has wireless, cable, and telephony. Sprint has a long-distance network and a wireless network and has recently acquired MMDS spectrum to allow it to access the home. Internationally, other big telecommunications players like BT and France Telecom also own multiple access techniques. So there are a significant number of players that could potentially drive this scenario.

In addition, there is a compelling logic to the notion that highly specialized systems will have to be developed in order to enable a truly broadband wireless future. Thus, the multiple system types will all have to prosper in order to deliver total solutions into the numerous environmental and usage profiles.

We have already shown that 5.4 GHz of additional spectrum beyond IMT-2000 will be required to support peak data rates of 200 Mbps. Acquisition of this amount of spectrum will inexorably force wireless systems into higher frequency bands. Given the current spectrum occupancy situation, we would suggest that this spectrum would most likely be found above the current LMDS band (i.e., above approximately 28 GHz). At these extremely high frequencies, propagation and fading limitations essentially rule out practical wide area coverage. Thus, we would expect that 200-Mbps wireless services would be provided by short-range nomadic systems. Coverage will be focused on specific hot-spot areas in which the density and requirements of end users can justify the costs of deployment.

Finally, end users will demand a solution that simplifies their user experience and provides control over the communication process while delivering near ubiquitous service. This demand could well provide the driving force for system integration if no single type is capable of meeting all of their significant needs.

The creation, however, of truly seamless networks that span multiple access systems provide a consistent "look and feel" and enable integrated billing is a daunting task, among many others. The entire wireless network paradigm will have to undergo a massive shift for this future to become a reality. Pulling off such a revolution will require levels of standardization, consolidation, and technical sophistication heretofore unheard of in the communication industry.

8.8.4 And the winner is ...

At this point the temptation is to sit back and simply wait for the future to arrive. We all fear that when it does, all of our scenarios and predictions are likely to look pretty naïve, if not downright ridiculous. This view, however, fails to take into account that the future is often created by those who have the audacity to believe that they can see it and the will to try to make it happen. Therefore, it's time to put our stake in the ground.

We believe that the most likely future is "integration wins." Although the obstacles facing this potential future are surely no less formidable than the other two, we ultimately believe that no single solution type will be capable of delivering the broadband wireless future. Therefore, operators will be forced to drive toward the integration of heterogeneous access systems into a single network. This future will not arrive easily. The power of hundreds of billions of dollars seeking ubiquitous delivery of seamless services at broadband rates in all environments will leave the industry no choice. The future is out there waiting to be created. Let's go get it!

8.9 A day in the life: 20 years out

His InfoPad chimes as Jeffrey Smith hurriedly crams down the last of his morning Pop Tart. This means one thing: The school bus is arriving and he is late, again. He grabs his backpack and runs out the door to the waiting bus. Enduring the usual abuse from the sixth graders in the back of the bus, he sits down next to his friend Jason. Jason has his Pad out and is interactively playing the latest Snork Wars game on it with some of the other kids on the bus. Jeffrey reflexively reaches into his backpack to join in and is hit simultaneously by two thoughts: one, that he has forgotten his new InfoPad, and two, that he's also forgotten to finish his English homework. He begs Jason to let him borrow his Pad and for a dollar, he does. Jeffrey connects to the school server and uploads his English paper, finishes outlining the chapter that's due and deposits it in the virtual homework inbox his teacher has set up. Done! He snickers to himself how he had justified his new InfoPad, the one that Jason already had, to his parents on the educational benefits, instead of the games he uses it for most of the time. How ironic! Now, back to Snork Wars. Maybe he can win back that dollar from that cheapskate Jason....

Kathy Smith, his mom, is also late—to the airport for an important meeting with Motorola, her biggest client. She scurries out the door, stopping only briefly to tell the home stereo to download the aria she's listening to into the car for her trip to the airport—she'd need it if traffic were the usual. In the car, the nav system displays an alternate route to the airport to avoid the worst of the traffic—it knows her destination from Kathy's calendar to which it automatically synchronizes as soon as it's turned on. Kathy relaxes for the drive as she listens to the end of the aria. She notices the familiar chime from her electronic personal agent—it suggests to her that, based on traffic and train schedules, she should park the car and catch the elevated train at Monroe Street to save 25 minutes on her trip. She decides to do it, but before she can act, her agent tells her she has an urgent call for which she needs to pull over. Kathy pulls off to the side of the road and takes the message—it's an indication that her father's medical alarm unit has gone off and that her agent could not reach him by telephone. She tries once again to call him herself and gives up. She instead has her agent contact her father's doctor. The doctor's agent replies that his office got the alarm, too, and that the ambulance had already been called. She silently prays that he's somehow okay and is glad that they'd had his medical alarm implanted last year, even though he'd objected to "that thing in my arm." Maybe he'd see things her way now—if he was okay. She tells her agent she'll have to cancel today's trip and to wait before calling her client—she'd try to schedule a remote presence session if possible later. She pulls away from the curb toward her father's house.

At her dad's house, Kathy meets the ambulance as her dad is being wheeled into the rear. She's relieved to hear that in fact he's okay—an irregular heart rhythm was detected and set off the alarm, nothing immediately life-threatening but nothing to ignore for long, either. They're taking him to Valley Hospital for observations and tests. She hugs him and can't believe that he'd forced his comm preferences to accept no calls at all for the morning—he'd "just wanted to get some peace and quiet for a change." She doesn't know whether to hug him again or choke him and decides on the hug. The ambulance wheels away, and she tells her dad she'd meet him later at Valley. Now that she knows her dad is okay, she'd better take care of her client. She puts her palm on her dad's lock, enters the house, and downloads her preferences into the house's comm system. She stands in front of her dad's live screen wall, a combination television, movie, and communication portal. She asks her personal agent to arrange a telepresence link with the client. She had really wanted to do this meeting in person, since her client contact was still new and she didn't quite feel comfortable negotiating with him over a remote link, but with today's situation, this would have to do. The agent complies and within minutes she's chatting with a life-size image of her client on the wall.

Back at home that night, her client meeting finished and her dad resting comfortably at Valley, Kathy catches her breath for the first time in about eight hours. She has just finished the last of her conversations with Dad's doctors, a specialist who'd tried to reach her earlier. Turns out he couldn't get through to her because he wasn't on her approved list, and she'd had privacy mode cranked up while at Dad's house talking with her client. She gets a snack and walks through the living room up the stairs to her bedroom, past Jeffrey, who's busy playing, reediting the latest Snork Wars movie, switching actors just for fun. She hopes he's done with his homework, but she's too tired to ask. In her room, she lays down next to her husband Joseph, who is logged into a live local politics discussion group, where a group of their neighbors are discussing prescription drug benefits with their state representative. Again. Kathy smirks and thinks that maybe they'll never get that one quite right. Jeffrey calls up from the living room on the comm screen and asks if she wants to play an interactive game with him. She decides to ignore the politics and take him up on it—he'll probably win yet again when she falls asleep from exhaustion. As she shoots at Jeff's ship, she thinks how funny it was that when she passed him in the living room was the first and only time she actually saw him today. She makes a mental note to play in person tomorrow and falls into a deep, well-needed sleep. It had been a very, very long day.

References

[1] Maxwell, K., *Residential Broadband*, New York: John Wiley & Sons, 1999.

[2] http://www.intel.com/intel/museum/25anniv/hof/moore.htm

[3] Sisson, P., "The New WTO Telecom Agreement," *Telecommunications*, September 1997, pp. 24–32.

[4] Washington, R., "SQUARE DEAL: Industry to Rate Radiation Output of Cell Phones," *Boston Herald*, July 19, 2000.

[5] Postrel, V., "Economic Scene: When It Comes to Enforcing Taste, It's Best to Tread Lightly—If at All," *New York Times*, July 13, 2000.

[6] Friedman, T., *The Lexus and the Olive Tree: Understanding Globalization*, New York: Anchor Books, 2000.

[7] Tapscott, D., *Growing Up Digital: The Rise of the Net Generation*, New York: McGraw-Hill, 1997.

Appendix 8A
Wide area cellular system considerations

8A.1 Spectrum considerations

The International Telecommunication Union has worked through its member organizations to do a thorough analysis of the spectrum required for third-generation cellular systems identified as IMT-2000.[2] They evaluated population densities in major metropolitan areas in North America, Europe, and Asia Pacific in urban, in-building, urban pedestrian, suburban, and vehicular environments. They also defined various types of services for IMT-2000 systems. These services were defined as follows:

▶ *Speech services:* This service is defined as a toll-quality voice service.

▶ *Simple messaging:* This service is defined as simple text messaging up to 14 Kbps.

▶ *Switched data:* This service is defined as circuit switch data at rates up to 64 Kbps.

▶ *Asymmetrical multimedia services:* This service is characterized by more data flowing to the terminal than from the terminal. Two types of services are being considered in this category:

  1. *High multimedia:* A user bit rate of 2,000 Kbps in one direction and 128 Kbps in the other direction;

  2. *Medium multimedia:* A user bit rate of 384 Kbps in one direction and 64 Kbps in the other direction.

▶ *Symmetrical multimedia services:* This category consists of voice and/or high-speed data and/or video and/or image services that require equal data rates to and from the terminal. Two types of symmetrical multimedia services are considered:

  1. A user bit rate of 128 Kbps in each direction;

  2. A user bit rate of 384 Kbps in each direction.

---

2. "Spectrum Requirements for IMT-2000," International Telecommunication Union, Radiocommunication Study Groups Document 8/80-6, April 26, 1999.

They also identified minimum practical cell sizes for the various environments of 80m in building, 630m for the urban pedestrian environment, and about 1,000m for the vehicular environment. They defined penetration rates, call durations, and activity factors for each of the services. These factors were then integrated into a model that resulted in the bit rate/cell required for each service in each region. This bit rate/cell was then translated into a spectrum requirement by identifying the net system capability of the IMT-2000 systems. Net system capability was chosen to be 0.15 bits/s/Hz/cell for data systems and 0.1 bits/s/Hz/cell for voice system for region 2. Slightly lower numbers were used for regions 1 and 3. The result indicated a need for approximately 400 MHz of spectrum for all these services.

This information can be used as a starting point for identifying future spectrum needs beyond IMT-2000 and noting when the requirement is so large as to not be viable. Looking 20 years out does require some change in the assumptions over the IMT-2000 calculations. For 20 years in the future, the new assumptions were made in the following areas:

‣ Penetration rates were assumed to have moved to 75% for all services.

‣ Net system capability was assumed to improve from 0.15–2.4 bits/s/Hz/sector over the 20-year time frame.

Using these assumptions, the spectrum required for medium multimedia, high multimedia, and highly interactive multimedia was calculated assuming a 10X and 100X increase in the bit rate required for each service. The result is the need for 250 MHz to provide a 10X increase in bit rate and 5.4 GHz additional spectrum required to provide a 100X increase in bit rate. It is highly unlikely that spectrum anywhere near the 5.4 GHz required for a 100X bit rate increase will be available in 20 years based on past trends for the rate of identifying and clearing spectrum for cellular services. Therefore, without a significant reduction in the spectrum required to provide 100X bit rate, it is unlikely that capability will become commercial in the 20-year time frame.

This spectrum limitation does not mean that 100X bit rate is unachievable. It does mean, however, that some significant improvements will be necessary in technology to achieve it. We have already been pretty aggressive in projecting that net system capability will increase from 0.15–2.4 bits/s/Hz/sector. So, while some additional improvement may be possible in net system capability, it is not likely that improvements will be enough to provide anything close to the 10X improvement required to get to a 100X bit rate. Another assumption that was made in the spectrum analysis was the minimum cell size. A simple analysis would indicate that we need to have cells with one-third the cell site spacing to move from the 10X bit rate to 100X bit rate through improvements in frequency reuse efficiency. However, one of the factors limiting the minimum cell size is the ability to acquire sites at closer spacing. Another factor is the change in the propagation environment as cells are deployed closer together in urban environments. These issues are related to the change in the propagation exponent causing less isolation between cells and therefore more interference. Another factor is the more variable scattering environment due to multiple scatters at or above the antenna height also resulting in more interference. These factors will need to be considered in any move to very small cells.

8A.2   System economics

The spectrum analysis was based on achieving 10X and 100X the bit rates over the IMT-2000 defined WCDMA system. The peak bit rate to the end user for IMT-2000 is 2 Mbps. Therefore, we need to understand the implications of peak bit rates of 20 and 200 Mbps in areas limited by the system range instead of the required reuse efficiency. Current systems, however, typically only provide voice coverage at approximately 10 Kbps. It is of interest then to look at the relationship between bit rate, the frequency of operation, and the cell radius. Table 8A.1 shows the relationship for a typical set of system parameters covering the range of bit rates of interest. The results are based on the Hata model and the reference is with respect to 800 MHz inbound. This reference is used since voice systems are typically inbound-limited, so their design is based on

Table 8A.1
Cells Required for 10-Kbps, 800-MHz Systems Versus End-User Bit Rate

Bit Rate (Mbps)	800 MHz	2 GHz	4 GHz
2	4X	10X	20X
20	10X	30X	60X
200	40X	100X	210X

achieving adequate inbound system gain. For these high bit rate systems, however, the inbound path will probably be allowed to degrade since higher rates are typically expected in the outbound direction. Therefore, while the reference is for an 800-MHz system at 10 Kbps in the inbound direction, Table 8A.1 compares the number of cells as limited by the outbound path for the high bit rate systems.

As we move from systems that provide coverage at 10 Kbps to ones providing 2 Mbps, the number of cells in a low population density area increases by a factor of 4. If the bit rate is increased to 20 Mbps, the number of cells required increases by a factor of 10X over a 10-Kbps system in low population density areas. If the bit rate is further increased to 200 Mbps, the number of cells required increases by a factor of 40X over a 10-Kbps system in low population density areas.

The need to move to higher frequencies further compounds this problem due to higher propagation losses. Increasing the frequency from 800 MHz to 2 GHz requires approximately 2X the number of cells in low population density areas. Moving from 800 MHz to 4 GHz requires approximately 5X the number of cells in low population density areas.

Of course, there are a number of underlying assumptions built into this analysis that impact the results. These assumptions also highlight opportunities to reduce the magnitude of the problem. The assumptions on the inbound and outbound path are independent and must be considered separately. Assumptions impacting the outbound path include the base station transmit power, the base transmit antenna gain, the tower height, the subscriber antenna gain, and the noise figure of the receiver. By placing the power amplifier at the top of the tower and increasing the base transmit power we can potentially improve the ERP by approximately 6 dB. By using an antenna array at the base station, we can further increase the transmit power. If the base antennas are used for transmitter diversity, there is also some potential improvement. Using multiple antennas at the subscriber in a diversity configuration can also buy some improvement. Finally, improving the subscriber antenna system provides opportunity for further improvements.

On the inbound direction, the options for improvement are limited by regulations on the allowable subscriber unit transmit power. More complex antenna structures and low noise amplifiers can be used to improve the sensitivity at the base station. These improvements could potentially improve the inbound system gain by about 5 dB compared to an optimized downlink. It is expected, however, that many applications will be asymmetrical, requiring a higher bit rate on the downlink than on the uplink. Based on typical base station and subscriber transmit powers, it is expected that the inbound path will be about 8 dB weaker than the outbound. This will require that the maximum bit rate on the uplink be one-sixth that on the downlink near the edge of coverage.

Looking at the impact of increasing the frequency of operation, there is a significant increase in propagation loss. Antennas of a given size, however, will improve at roughly 10 log(frequency), so some of the propagation loss is offset by improved antenna gain.

We will assume that the system design will be based on the downlink and that the uplink will be degraded as the subscriber moves away from a site. Given the above assumptions, overall how much might we reasonably improve the system? We would guess a reasonable limit might lie in the 10–12 dB range at a given frequency. With these improvements factored in, the increase in the number of cells required in low population density areas is as shown in Table 8A.2. These higher bit rate systems then will require significantly more cells than current systems in areas of low population density.

Table 8A.2
Cells Required for 10-Kbps, 800-MHz Systems Versus End-User Bit Rate

Bit Rate (Mbps)	800 MHz	2 GHz	4 GHz
2	1.1X	3X	6X
20	3X	8X	16X
200	10X	30X	60X

CHAPTER

9

Contents

A bold view of the future

Michel Mouly

9.1 Introduction

When William Webb asked me to write something on the future of cellular systems, I started writing lengthy and detailed arguments analyzing the issue in all directions. I realized that the result was boring and figured that it would likely duplicate material in the rest of the book. So, between bland, carefully laid, and likely ideas, and daring, iconoclastic, more risky views, I chose the latter. Any attempt to predict the future *is* risky. So, please take the following for what it is, something aimed at stimulating your thinking, a challenge to what appears to be unquestioned axioms.

I have chosen to discuss three topics, which are somewhat intermingled. Short term comes first, with a discussion on the data services that are the driving forces behind GPRS and UMTS. The second thesis is the trend toward "centrifugism," my term for the provision of services from the periphery of the system rather than from its center.

Third is an analysis of the changes in the role that operators might play in the future.

9.2   Cellular data

Up to early 2000 the media were full of the brilliant future of mobile access to data services, and more precisely of mobile access to Internet. This was justified by the announced arrival of UMTS, with its astronomically expensive licenses, and because it appears as the logical convergence of the two major trends of telecommunications during the 1990s: the unexpected growth of cellular telecommunications and the rise of the Web. Combining the two areas seems the obvious road to the future, and work in that direction started in GSM and other second-generation systems many years ago. These efforts begin to be visible with WAP, with the delayed but always soon-to-be-launched GPRS, and, more recently, with announcements of UMTS. In mid-2000, some negative notes could be heard, which were linked to the disappointment of users with WAP services. The media always ended up optimistically, invoking the immaturity of the current products and the improvements to be expected from GPRS and UMTS.

In my view, the vision of widespread cellular access to Internet occurring in the near term is a simplistic idea, based on a superficial analysis, and missing a number of important points. Taking these into account, not only could the disappointment of WAP users have been predicted, but also the expectation of an even bigger disappointment with the services to come in the next few years. The evidence leading in that direction is plentiful.

The main and simplest point is the available data rate. Cellular access, contrary to wireline access, suffers from a very narrow bottleneck, the radio access. With existing techniques, a factor of at least 10 or 30 in bits/s exists between the bit rates achievable for the mass-market with a wireline access (e.g., ADSL) and those attainable with cellular access.[1] There's no clear reason why this gap should decrease in the next few years; on the contrary, there is ample room for improvements of wireline access! This gap is, and will be, the main source of disappointment for users. They are used to cheap and high rate wireline access, and they

---

1. Although ADSL is slow to roll out and expensive in some areas.

expect (with the help of marketing and media) something similar when they access through a cellular system, but they will not get it.

The weakness of cellular systems in delivering high data rates is partly hidden by the rather misleading habit of quoting peak rates rather than average rates. Usually, wireline access resources are not shared between several users, and their full capacity is available to each user. When an ADSL modem is specified at 2 Mbps, the user gets this entire capacity for themselves. Users do not notice a reduction of 10% or 20% of the capacity, but they do for more drastic reductions. On the other hand, when UMTS claims 2 Mbps, this rate will (if it is ever actually offered) be shared between many users. In practice, this means that some users might perceive a rate approaching 2 Mbps for short data blocks, but they will find that the larger the amount of data to be transmitted (e.g., file transfer, big Web pages, in fact anything big enough for the user to be concerned with the rate), the more the need to share the access, reducing the rate experienced possibly down to very small values when there is congestion. Curiously, cellular access is a self-defeating system: the higher the number of users, the worse the service. This can be said of the Internet at large. However, congestion in the backbone (i.e., not in the access part) can be (and is) overcome by the growth of the equipment. This is much more difficult for cellular access, with every percentage gained in spectral efficiency—and therefore in capacity—being more expensive, not so much in electronics but in site finding and site installation. Unfortunately, it is not obvious how these costs will be recovered. This takes us to the next point: the interest of the operators.

When extant services are compared in terms of revenue per unit of radio spectrum, the short message service (SMS) of GSM comes first, then speech. Other data services get less revenue per spectrum occupancy than speech. Worse, it is quite likely that the higher the average data rate, and the more the Internet is targeted, the less operators can tariff the actual spectrum usage. Also, the part of the revenue of Internet services coming from advertisement is limited by the poor display in the user equipment and might not reach cellular operators' pockets at all. Then, and because the speech services are far from saturated (it suffices to look to tariffs to see that), the economically sensible approach for operators is to extend speech and SMS, not data. So why so much hype about those services? In my view, it is simply because the problems are too technical for the media and because any operator defending publicly the thesis developed above would appear backward in the midst of the Internet revolution.

A third point backing the idea that cellular data services as imagined in GPRS and UMTS will not be a rapidly accepted by the mass market is the user interface. The current success of cellular is based on speech services and so is not hindered by this aspect. The local man-machine interface for basic functions is rather simple. The end-to-end interface is a man-man interface based on speech and does not require much in terms of local support. The consequence is the ability to provide simple, low-cost and low-weight, handheld user equipment.

On the other hand, Internet services are man-machine-man services, and the local man-machine interface must support an interface between the user and a distant machine. The current solution, adapted to fixed or semifixed access, is a computer, with big display, a keyboard, and a mouse (not taking into account all the software to go with it). To reproduce the facility offered by such equipment to a mobile user, in the same conditions in which he/she can efficiently use speech services, does not seem to be possible in the next few years. Such a revolution can (will?) come later, with such techniques as direct writing on the retina (to replace the screen) and detection of hand movements (to replace the keyboard and mouse). In my view, such a drastic revolution of the man-machine interface is required before mobile Internet access meets a level of success equivalent to that of mobile speech. A precedent is the Internet itself: The network existed and was used by a minority on the basis of a text interface, years before it started to have a mass-market success. This success came with hypertext markup language (HTML) and the Web, and with rather low-cost personal computers, (i.e., when a revolution happened in the man-machine interface).

A first conclusion, then, is that the heralded success of mobile Internet will not come soon. The market needs will be provided differently, probably with easily accessible pervasive wireline access, or short-distance radio access. True continuous coverage of high-speed access may happen someday but will require technical breakthroughs not yet announced.

The evolution of telecommunications in general, and the attempt to integrate Internet access to cellular systems, will have important side effects, and this takes us to the next of our topics.

9.3 "Centrifugism" of services

The second topic analyzes a trend that started 20 years ago and that has not yet reached maturity in telecommunications. Up to the advent of

Minitel[2] in France, the user equipment was passive, and all the added value was provided by the infrastructure of the telecommunications operator. Minitel was the first example of the more complex user equipment required to access commercial telecommunications services (for example, it has to be plugged to mains). It was still "dumb": It was only an improved man-machine interface, with control (e.g., echo) within the infrastructure. Cellular networks necessarily entailed the use of complex handsets, and with them appeared telephone handsets with screens, providing such simple and long-awaited added service as displaying the number when entered. Nowadays, GSM handsets provide an array of services locally, without involvement of the infrastructure (e.g., personal directory of abbreviated numbers).

The biggest revolution illustrating this trend is the Internet, which requires substantial local software to access services and to provide local services (e.g., browsers). Currently, the services provided by the Internet backbone are quite limited; most of the added value is at the periphery of the network in servers. In general, the service perceived by the user combines contributions from the local equipment and from distant servers.

This is part of a trend moving the added value out from the center of the network and toward the periphery. In my view, this trend will be pursued, until most, if not all, services besides information transport are provided by the user equipment and by servers independent from the transport networks. Cellular systems appear as the best place to operate this revolution because user terminals already have a high level of complexity due to the access technique. Specific attempts in this direction include CAMEL[3] and the SIM toolkit,[4] as well as attempts to integrate Internet access, as illustrated by WAP.

In practice, such an approach impacts the way in which value-added services are designed, on user equipment, and on the role of the operators. The latter topic I will address on its own. We will see more and more services designed as a combination of a user equipment feature and of distant services. This requires cooperation between user equipment manufacturers and added-value service providers. Techniques will evolve so that this cooperation can be done independently of the transport operators.

---

2. Minitel is the name of a system provided by France Telecom that utilized small terminals connected to the telephone line. The terminals had a keyboard and small screen. They could be used to retrieve selected information such as the expected arrival time of flights. In a way, Minitel was a forerunner to the Internet.

An important evolution in user equipment that should come soon is the possibility of personalizing it by adding software or physical add-ons to adapt it to the services, local or distant, as required by the user. Downloading functions is something that is natural in personal computers and in accessing Internet services, but it is new in pure telecommunications equipment. User equipment will become more and more like a small network, with parts specific to access, parts specific to subscription (for example, the SIM), and parts specific to services. The possibility of combining several access modes in the same user equipment, sharing the subscriber and services parts, offers an alternative to the all-powerful cellular system, albeit one difficult for many to envision at present. For this to come, another revolution is needed, in the role of operators.

9.4 The operator role model

The concept of role model is not much advertised, though it is a critical aspect of telecommunications. It deals with the economic roles of the different actors as seen by the subscribers, and with the relationships between them. It deals with such ideas as subscription, roaming, invoices, and refunding the different companies involved in service provision.

A few decades ago, a single role model could be found in telecommunications, based on state monopoly. The relationships were limited to that between the operator (the administration) and the subscriber, and the relationship between administrations of different countries, to deal with international calls.

In the 1980s, variations started to appear: with the breakup of AT&T in the United States, with the concept of roaming in Scandinavian

---

3. Customized Applications for Mobile Networks Enhanced Logic (CAMEL) is a feature introduced in GSM to enable the support of operator-specific services for customers even when they are roaming. It consists of applying an intelligent network architecture and standardizing generic exchanges of data between the home network of a subscriber and the network in which it is located in order for the former to dictate personalized behavior to the latter.

4. The SIM toolkit is a set of generic mechanisms enabling an operator to offer personalized services to its subscribers even when they are roaming by letting the SIM "control" the behavior of the mobile equipment. It is to the subscriber equipment what CAMEL is to the network.

Europe, and also with the onset of "deregulation"—GSM was launched entirely on the basis of independent operators providing access to backbone networks. The Internet introduced other, very important novelties, in particular the distinction between access, transport, and services (with the result that in many cases one has to pay three different operators for some services: for the access, the ISP, and the true provider of the service). Internet also opened new opportunities for access (as opposed to long-distance transport), for instance, to cable operators or electricity providers.

The current role model, as applied in cellular services, is a mixture of the different influences and does not appear to be stable. The evolution will still go on, and it is interesting to conjecture on the form of the final model.

Three main roles can be distinguished: the transport operators, providing transport of information between distant points; service operators, providing value-added services; and subscription operators, providing interface with subscribers, in particular for the financial aspects. Transport operators can be further split between access (e.g., cellular) and long distance. So far, the organization of telecommunications has been to a large extent "vertical": Operators are playing all roles, each one around one access technique. Cellular operators, which operate cellular access systems, also handle subscriptions (especially in the case of roaming) and provide value-added services (e.g., voice-mail system). In parallel, wireline access operators fulfill the same functions for wireline access, and most often they offer long-distance transport in addition. Internet is also an area where some specialization is the rule for subscription operators (ISPs) and for a constellation of service operators. This verticalization leads to specialized user equipment (hardware and software), specialized subscriptions, and often specialized services.

Now, let us see the issue through the eyes of the subscriber. The vertical approach is more a source of limitations than anything else. Users don't really care about the underlying access techniques, and they do not see any advantage in having different calling numbers or different invoices according to the access mode used. On the contrary, the ideal situation would be a single subscription and user equipment able to support and to choose the best available access technique and transport network. Steps in that direction have been taken in the cellular world, with multimode handsets, including cases of terrestrial access combined with satellite access. This could be extended further, to include wireline access or local area network access.

Such an evolution is possible technically but requires a drastic change in the organization of the operator. A role model adapted Such an evolution is possible technically but requires a drastic change in the organization of the operator. A role model adapted to such an approach consists of separating clearly the subscription operators from the transport operators. This is roaming pushed to its extreme: The "home" operator deals only with subscriptions and plays the role of financial intermediary with the "visited" (transport) operators, including responsibility for fraud. In parallel, services must be split from transport techniques so that the same service can be accessed by different means.

Seeds for this evolution are already in place: The concept of the subscription operator[5] is a logical consequence of roaming in the cellular world and an extension of the ISP in the Internet world. Distinct service operators are already the norm in the Internet, and some examples can be found in cellular systems. The evolution, however, will probably be slow for various reasons. Technically, the roaming method put in place in GSM and continued in UMTS or in some satellite systems is not accepted outside that circle; for example, the Internet is developing its own model under the name of mobile IP. Another example is the reluctance of U.S. operators to use smart cards. It will take some time to obtain convergence, including with wireline access. Another factor is that established operators are reluctant to be reduced to the role of transport operators. Still another factor is the licensing approach applied by regulatory authorities. For the moment, it encourages a vertical approach and does not distinguish between licenses for deploying access networks (e.g., spectrum usage) and licenses for managing subscribers (e.g., number prefixes). On the other hand, it can be expected that the market and the technical evolution of handsets will provide sufficient pressure for the evolution to take place, certainly within the next 20 years.

Interestingly, this implies the disappearance of cellular networks as we know them today. They will be reduced to their radio access network part and will be only a small component of a wider system along with various wireline accesses, short-range radio access, and satellite access.

---

5. Sometimes termed "virtual operator."

9.5 Overall conclusion

In a few short sentences, my views of the future are these. The cellular speech services will continue their growth for many years, for at least a decade if looked at on a global scale. Data services will not have the initial success currently advertised; they will start slowly, in countries where "high-tech" is an important sales argument and where speech services enter saturation. Exceptions might be specific services: for example, machine-machine services, and services like SMS or e-mail—that is to say, very low data rate data services, presented as a complement to speech.

Mass-market mobile data services might have some success later, say around 2005–2010, depending on technical revolutions that yield real benefits—which are not yet embodied within UMTS—both in the area of spectrum usage and of man-machine interfaces. Mobile access, however, will remain far behind wireline or short-distance radio access, which will become the preferred access means to such services as the Internet. This evolution will be accompanied, or possibly preceded, by a continuation of the revolution in the role of operators and a corresponding adaptation of the user equipment. Large telecommunications operators will be reduced to providing transport infrastructures. Each user will have a single subscription (for example, with his or her banker), independent of the access modes he or she uses and will access a multitude of services, involving distant servers and software local to his or her equipment, thanks to a variety of access modes, cellular being just one of them.

CHAPTER

10

Contents

Realizing the mobile information society

Tero Ojanperä

10.1 Introduction

The growth in the penetration of mobile phones has been tremendous—so tremendous that to many people wireless communications equals mobile phones. Soon, practically every person in developed countries will have a mobile phone—from school children to elderly people. Voice has gone wireless. Given this ubiquity, what can be new be in wireless communications? The next 20 years could be divided into three phases:

1. *Phase 1:* Voice increasingly becomes wireless, 2000–2005;

2. *Phase 2:* Development of mobile information society,[1] 2006–2010;

---

1. The term "information society" has been much used. Here we mean the capability of users to access any information that they want (within reason) wherever they are. This capability will likely lead to social changes, such as increased home working, and hence the term "information society."

233

3. *Phase 3:* Full realization of the mobile information society, 2010–2020.

During the first phase, mobile penetration reaches its final peak, which is approximately 100% in the most developed countries.[2] The mobile phone has truly become the preferred phone, and the number of mobile users has exceeded the number of fixed phone users.

In the second phase, wireless data services will emerge in the mass market. Mobile users will be able to access all types of information whenever they want, wherever they want. Today's primitive and little-used data services such as wireless application protocol (WAP), iMode, and location-based services become widely used. Mobile devices in all types of form factors emerge as the primary means to access the Internet, and totally new services for mobile users are introduced.

In the third phase, seamless interaction with humans and machines through wireless devices using voice, sight, and touch will become reality. Integrated mobile devices will communicate our feelings and thoughts to other people or receive information about our environment, wherever we happen to be. Today's separate communication networks will evolve into one single, distributed network spanning the whole globe and offering the same type of service set regardless of the air interface used.

One of the key drivers for future development is the need for growth among the operators and manufacturers in the wireless industry. Wireless data is required to maintain growth of revenue for mobile operators. The growth of voice-only services will continue with saturating penetration figures once nearly everyone owns a mobile phone. The new business opportunities offered by the mobile Internet are currently driving the investments and new business innovations that will lead to the developments of the future.

10.1.1 Requirements for future communications systems

In this section, we explore the requirements for future wireless communications. This will be examined from three angles: user behavior, services, and user devices.

---

2. This had already occurred, for example, in Finland during 2000.

10.1.2 User behavior

Mobile phones have freed the user from physical location to make calls. This is reflected in the very natural way that humans prefer to communicate with each other regardless of the place where they happen to be at a specific moment. Voice is only one type of communications. Vision is another, and it is already emerging in wireless communications through picture messaging.

Other natural needs for humans are the need to organize things and the need to have fun. Things that make this possible to carry out anywhere have latent demand. How many times have you wished to do something useful or entertaining while waiting in line, sitting in a taxi, or commuting in a train or bus?

A newborn baby has zero tolerance and requires immediate attention and fulfillment of his needs. Although adults are more sophisticated in behavior, instant gratification is a fundamental desire for all people (i.e., all information that one wants should be delivered immediately). When you hear about the latest music hit from your friend, you would like to download it immediately to your own terminal without waiting several minutes—and how many times have you felt frustrated trying to download your e-mail while on the road? With all the attached information it takes too long.

Hence, the answer to the question of how much bandwidth is required in the end-user terminal can be derived from the user's behavior. The need for instant gratification and zero tolerance means the answer is "as much as can be delivered." For example, a user browsing a travel agent's Web pages wants the three-dimensional picture of a sunny beach delivered immediately, not after 10 seconds. It is more difficult to solve what makes economic sense.

10.1.3 Services

Describing and predicting future services is very difficult. Assume that five years ago a market study was conducted where people were asked if they would like to send text messages of up to 160 characters through their fixed phones. Most likely, nobody would have liked this type of service. SMS, however, currently accounts for approximately 10% of some operators' revenue.

Future wireless communications offers wide variety of services including the following:

- Conversational, interactive services (voice and video);

- Location-based services;

- Games and entertainment;

- Messaging services;

- Mobile Internet.

Even though voice is already now a saturated service in many markets, there might be new developments that will further accelerate the migration from fixed to mobile networks. Wideband codecs using double the sampling rate of normal voice codecs are currently in the standardization process. This means that instead of normal 3.6-kHz voice band used in all fixed networks and in mobile networks, a 7.2-kHz sampling rate is used. These codecs will deliver clearer sound than traditional codecs, and therefore could bring the voice quality of mobile networks to above that of fixed networks.

Wireless video is one of the most touted new services for third-generation mobile systems. It should be remembered, however, that the video phone was first introduced in 1927 by AT&T and has often since been predicted as the next major commercial success. In the mobile world, it might finally succeed since it fulfills one of the basic needs of human beings, to express feelings using all senses. Not only can you deliver your own picture, which might not be the most interesting part, but also the picture of your environment, perhaps a nice beach while on holiday or a repairman sending an image of a machine that he is working on to the service center for further instructions.

Location-based services make use of the location of the mobile terminal. For example, your mobile terminal can locate where you are and then deliver information that is relevant to your current location such as a local map or restaurants in the neighborhood.

Games and entertainment will most likely account for the majority of future wireless services as they do in the fixed Internet. The personal nature of the mobile phone and the fact that people usually carry it with them makes it ideal to deliver entertainment services. The first network games for mobile users are already emerging.

SMS has been widely successful in some markets, for example, in Scandinavia. Applications of SMS include delivery of different types of information such as weather and stocks. It is also possible to buy a can of Coke by dialing a certain number or sending a short message from your mobile to a given number. You can also pay for a car wash using your mobile phone. This is a simple but very tangible example of mobile commerce (m-commerce)—a new way of purchasing. Picture messaging and multimedia messaging are the next steps in the messaging services.

The mobile Internet enables access to information regardless of location. This is an advantage that should be the basis for service development because those services that are driven by the development of wireline networks face the problem of bandwidth in the wireless networks. In the fixed network, for example, DSL technology has been a significant breakthrough in recent years, and, consequently, broadband access to the Internet has become a reality for those consumers able to receive it and willing to pay the premium often charged. This has enabled, for example, high-resolution, three-dimensional graphics to be delivered into the home. Mobile Internet will face problems in trying to match the wireline data rates and service level. Therefore, the unique advantage of location information should be utilized.

10.1.4 User devices

Today, we have primarily one type of form factor for mobile devices. Moreover, mobile phones have been referred to as clumsy devices to access the Internet. Small, black and white screens and tiny keyboards have made it difficult to implement really fancy applications.

This picture will change dramatically in the future. Miniaturization of the core components such as chips and radio frequency parts will enable these to be built into almost any form factors, as opposed to the current situation where the form factor is somewhat limited by the need to package all the components that comprise a cell phone. Already, wristwatch mobile phones are a reality. High-resolution color displays that can even show a movie will become a reality in the near future. Looking forward to the next 20 years, it is not hard to imagine that there will be 3-D displays in mobile devices.

Completely new types of user devices and user interfaces will emerge in the form of wearable computers. For example, the screen will be integrated into eyeglasses and user commands entered using a voice recognition system. It will become possible to create virtual reality around the user and through this communicate with other users as if they were present. There will be many attempts to find the right form factor for different types of terminals such as videophones, Internet access devices, and wireless game consoles. In particular, the next few years when third-generation networks and wireless data will emerge will be a very fruitful time for all kind of trials with new types of form factors. After the trial-and-error period, however, for each device category, a de facto standard will emerge, as has happened for almost all consumer electronic devices.

Currently, we carry a number of various devices for different purposes with us: for example, a mobile phone, a digital camera, an MP3 player, and an electronic calendar. The convergence of devices, however, is changing all this. In the future, our personal trusted device will contain elements from all of these devices enabling faster, more effective, and easier ways of doing business.

So far, computational power in the mobile phones has been limited. This might be a problem for applications such as games. In the near term, a solution could be to locate computational power in the network and only the results are delivered to the mobile device. In the longer term, more efficient computers will emerge and could solve the problem. For example, quantum computers with computing power of thousands of times more than today's fastest computers could be reality within the next 20 years.

10.2 Air interface development

The need for more bandwidth and the cost pressures calling for higher spectrum efficiency drives the development of new air interface technologies. The higher the throughput in the network, the better from the end user's perspective. On the other hand, the spectrum efficiency determines how efficiently the bandwidth can be delivered to the end user and is of primary concern for the mobile operator.

To increase the throughput, existing air interfaces will be stretched as far as is technically and economically possible. Figure 10.1 depicts this development. Common features for all of these developments are the evolution of the standard using a change of modulation technique. Limiting factors have been receiver complexity and transmitter linearity requirements. Mobile systems have three standard carrier bandwidths: 200 kHz, 1.25 MHz, and 5 MHz, using time division multiple access (TDMA) and code division multiple access (CDMA) techniques.

The US-TDMA system, also known as digital advanced mobile phone system (DAMPS) and interim standard (IS) IS-136, first evolved within a 30-kHz carrier bandwidth using multislot and higher order modulation

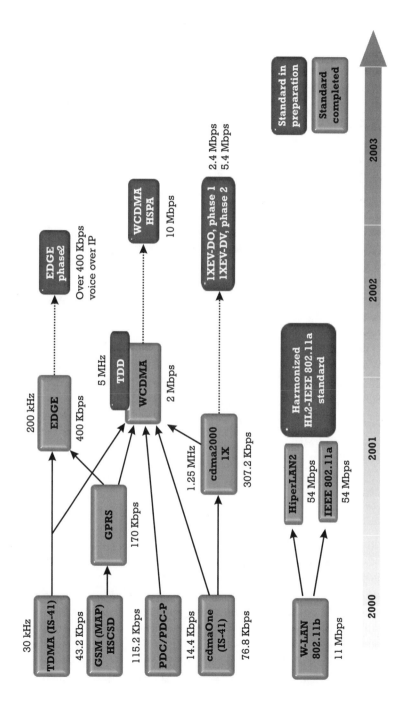

Figure 10.1   Evolution of mobile radio standards.

(8-PSK—phase shift keying) techniques. After the evolution of the 30-kHz channel became too difficult, the US-TDMA community selected EDGE as its future air interface evolution. GSM also used multislot techniques [high-speed, circuit-switched data (HSCSD) and general packet radio service (GPRS)] and higher order modulation (8-PSK, EDGE) to evolve the 200-kHz carrier. The next step could be multicarrier EDGE.

The main third-generation standard is wideband CDMA, commonly known as WCDMA. WCDMA high-speed packet access (HSPA), intended to be used in the downlink, uses higher order modulation to increase the maximum data rate from 2 Mbps to 10 Mbps. WCDMA time division duplex (TDD) will be used in the so-called unpaired frequency bands allocated to third-generation networks.

The IS-95A standard, also known as cdmaOne, evolved into IS-95B, which uses parallel CDMA codes to create high-speed data services. The next step in the IS-95 evolution is cdma2000 1X (1X means that the carrier bandwidth is the same as in the IS-95 standard, namely, 1.25 MHz), which adds coherent detection into the uplink in order to increase radio performance and higher data rates. 1X-EV (evolution), in addition to parallel CDMA codes, uses higher order modulation techniques to further increase the maximum data rates to 5 Mbps.

Among the various W-LAN proposals, the IEEE 802.11b standard, originally providing 2-Mbps and now 11-Mbps data rate, has emerged as the mainstream standard in the 2.4-GHz frequency band. It is evolving further to support even higher data rates up to approximately 20 Mbps. In the 5-GHz frequency band, IEEE 802.11a and HiperLAN2 standards provide a maximum data rate of 54 Mbps, and this will be pushed higher.

BlueTooth provides short-range wireless connectivity and can be positioned as a cable replacement. It will be used to connect different devices into the mobile phone. There are also studies that investigate the possibility to use BlueTooth as a public access method. The next-generation BlueTooth, which is currently under development, will provide higher data rates than the existing 700 Kbps.

In addition to the above-mentioned mainstream techniques, there is a large number of different types of air interface proposals and products that are variations from the above-mentioned schemes, as well as new types of air interfaces. One very interesting new scheme is ultra-wide band (UWB), also called impulse radio, which sends very short impulses and thus spreads the information over a very large bandwidth on the order of several hundreds of megahertz up to few gigahertz. It has been proposed that UWB systems could be overlaid with current radio systems.

At the dawn of deployment of the third-generation networks, the research community has started to discuss fourth-generation networks. Key targets are:

‣ Higher data rates without compromising cell range (i.e., coverage);

‣ Increased spectrum efficiency.

It is questionable, however, whether a new air interface would be required to achieve these targets. For example, the WCDMA air interface can be stretched toward this goal with new innovations.

Radio is only about 100 years old. At the time of its invention, who could have predicted that a major part of mankind would use it by 2000? Given the accelerated technology and innovation cycle, we predict that totally new technology for communications, not the evolution of the current radio technology, will be invented within the next 20 years.

10.3 Network development

Cost and flexibility are drivers for transition of current circuit-switched mobile networks into packet-switched, IP-based networks. Until today, there have been separate networks for voice and data; digital convergence, however, will create only one IP-based network to provide all these services. This development will bring IP, and the end-to-end philosophy and modularity, into the telecommunications world. The new generation of IP protocol, IPv6, will be essential in implementing the next-generation mobile networks. IPv6 provides huge IP address space, excellent scalability, and integrated security support. The challenge in developing the IP-based networks is how to bring the telecommunications-like functionality (for example, reliability) into the IP world.

The ultimate goal is to be able connect all the types of radio access networks, both WANs such as EDGE and WCDMA and LANs such as BlueTooth and W-LAN, into a common all-IP network. Scalability of the network solution is essential to implement this goal.

How many times have you thought, "If I only had performed backup of my essential information in time," after having lost hours of work while working with a personal computer? Similar problems will start to emerge in mobile devices when they are transformed into personal trusted devices. Digital convergence will put more and more functions into the mobile device. What happens if your mobile terminal gets lost or it breaks down? The key, of course, is to have a synchronized version of your information stored in another place. Also, the fact that there will be lots of information in your device that might be useful for other people makes it important to have it up to date. Current synchronization schemes are manual and require specific action from the user. Synchronization of user information should not require user attention but should be automatic. Technologies to do this will emerge in the near future. This will also create lots of machine-to-machine communications in the network in the form of exchange of information between different databases.

With the Internet model everything becomes addressable, and a user's information can be located in different places in the network. Therefore, the importance of security will increase dramatically. Guaranteeing that your own information remains private and inaccessible to other users without your permission will be critical. Without solving the security problem, the development of mobile information society will be much slower than predicted.

10.4 Future scenarios

We predict that the future of wireless communications will be centered around one of the following two scenarios:

> ‣ Cellular dominance with further evolution;

> ‣ A distributed network paradigm.

In the first scenario the traditional cellular paradigm dominates. Networks will be built starting from wide area coverage, and maximizing bit rates and coverage at the same time is desired. Transition from second-generation to third-generation networks (from GSM, US-TDMA, and EDGE to WCDMA) will happen within 10 years from the launch of 3G networks. Drivers for this transition will be better service capabilities and spectrum efficiency. Second-generation networks, however, will still have a significant number of cellular subscribers by 2010. Complementing technologies such as W-LANs will be integrated into cellular networks. A seamless service offering is driving this scenario.

In the second scenario new technologies such W-LANs, wireless mesh networks, ultrawide-band radio, or some new radio technology emerge as separate networks and obtain mass-market scale. Some new service tailored for these new networks might drive this scenario. Initially, networks will emerge as islands of coverage. Imagine that each electronic device could be a wireless router connecting it into surrounding nodes. Wireless devices can connect into these routers. Gradually, the density of the network will increase and roaming capabilities will emerge. Understanding how to make a worldwide network based on this concept and be able to provide adequate quality of service remains a key challenge. The demand for high bandwidth could drive this second scenario. Traditionally built, hierarchical networks could ultimately be too limited to support huge bandwidth requirements in all the places that users want them. A more distributed network structure is required.

The book *The Innovator's Dilemma* [1] describes the development of mills, mainframe computers, and cranes. In all cases, a distributed, smaller architecture took over the earlier paradigm of centralized, "the bigger, the better" architecture. This would support the latter distributed network architecture paradigm.

10.5 Concluding remarks

In this chapter we have explored the future of wireless communications. Human nature, characterized by zero tolerance and the need for instant gratification, is driving the development of wireless communication systems to add more functionality and higher bit rates. User devices will enable communications using more senses: hearing, vision, and perhaps touch. New services, not possible to predict today, will emerge, encouraged by open platforms provided by IP-based networks. Air interface development will be based on the evolution of currently known waveform structures. Future network architectures will be centered on two scenarios: hierarchical versus distributed, where the distributed architecture will emerge as the dominant one. Finally, as the most exciting vision, within the next 20 years a new technology for communications, not based on currently known radio, will emerge. This is something that is as impossible to imagine today as the current use of mobile phones was 100 years ago.

Reference

[1]   Christensen, C. M., *The Innovator's Dilemma*, Cambridge, MA: Harvard Business School Press, 1997.

CHAPTER

11

Contents

Learning from the past

Malcolm Oliphant

11.1 Introduction

Cellular systems have evolved through several generations characterized by their salient technologies: analog and digital. They are customarily numbered as first generation for analog, second generation for digital, and third generation for enhanced digital. Each generation moved the industry into a more advanced stage of a wireless evolution. The first-generation systems were reasonable alternatives to wireline service if the latter was not available, or if the user had to be mobile. The second-generation digital systems offered services and features that were complementary to wireline services in some markets and were viable substitutes to wireline in more mature markets. Today's third-generation initiatives seek to drive the mature markets into saturated ones in which mobile terminals are everywhere. If we assume the early years of 3G achieve this saturation, then we have to ask what comes next. How can there be sustained growth in a saturated market of ubiquitous wireless mobility?

This chapter suggests that aggregating large numbers of different kinds of networks into an enormous number of virtual PANs will provide the means for sustained growth beyond the 3G era. The only missing element will be the *content*, which will be very different from that likely to appear in 3G systems. We start by transporting ourselves into the future, to 2018 where we meet Peter, who uses hundreds of wireless applications every day. Although some of the applications are purely wireless ones, most extend into many kinds of networks as they follow Peter around and show their presence in different ways. Multiple applications typically work together to embed Peter in a virtual personal area network that changes its character according to Peter's location and his circumstances. Peter's PAN sometimes breaks up into little islands in which appliances and devices, most which belong to Peter, communicate among themselves on Peter's behalf. Enabling devices to talk to each other is what allows even more wireless terminals to be sold, and the rich tapestry of appliances and circumstances that come and go in adaptive personal areas gives the necessary fertilizer for a steady stream of applications. Peter is generally not aware of the PAN that surrounds him and seldom interacts with the applications explicitly, but he is so dependent on them that it is difficult for him to imagine living without them.

Having seen a glimpse of what the post-3G world could look like, we take a look back to the past, back to the precellular world, to see how we could move from the emerging 3G WANs of today to the hundreds of millions of PANs that could invade them tomorrow.

11.2 Peter in 2018

Because it is already 2018, it is easy to imagine someone with a BlueTooth-enabled home[1] in Albuquerque, New Mexico. A wireless user only 22 years old whose name is Peter (as in Peter Pan), lives in a post-3G environment and uses his wireless handset almost everywhere. His terminal is actually a sophisticated combination of a wireless communications terminal, personal digital assistant (PDA), and personal information manager (PIM), which Peter uses in the office, at home, and while

---

1. For a description of BlueTooth, see Section 3.2.5.

traveling; he takes it everywhere. This particular terminal conforms to an outdoor mobile radio protocol, and another indoor high-bandwidth radio protocol. Peter decided to get the broadband hookup option, which lets him connect his terminal to special connectors found in most homes and apartments, hotel rooms, and airport kiosks all over the world. The connections provide high-speed communications for most of the multimedia applications that he uses regularly. Peter bought the same terminal he got for his grandmother more than two years ago. His grandmother's terminal, however, is fitted with the emergency medical option. This is a service that automatically calls emergency medical personnel in the event of stroke, heart attack, or other medical emergency that can be cured if special care is administered within two hours of the onset of problems. Because the indications of the problems can be anything from unnoticeable to dramatic, sensors, which are sensitive to the life-threatening conditions, were developed for people to wear around the neck or wrist, like jewelry. The sensor's presence is discovered by an appropriately equipped wireless terminal through a local RF network that consists of the sensor appliance itself and the wireless terminal. Upon receiving an alert from the sensor, the wireless terminal calls the nearest emergency medical team regardless of where the victim may be. The local network determines the victim's location and adds the location details to the call for help that goes out to the local emergency medical team. The system of fast and accurate notification and locating coupled with very effective medical intervention has been accepted by society as cures for many chronic medical conditions. Peter's grandmother is still alive today thanks to this system. Just about everyone over 60 years old has one these days.

Peter has an account with one of the local communications providers, which set up a VPN profile for him. Peter's VPN profile is specific to his multimode terminal, his fixed connection at home, and his general communications needs and travel habits he declared when he set up his account two years ago. His VPN also includes provisions for some interworking functions with his employer's wireless LAN at work. Peter's communications provider also set up a residential PAN as part of his VPN profile. The PAN includes a personal base station (PBS), which works with the high-bandwidth indoor radio protocol in his terminal, the fixed access ports in his home, a wireless network of all his home appliances, and some small TV cameras placed throughout his home. Peter lives alone, and because he travels a great deal (he's an editor for some foreign language publications), he likes the convenience of being able to look around inside his home while he's away.

Peter is at work now, but he's decided to go home early because he wants to prepare some schedules for a short visit he is having with a college classmate who is coming to town for a few days to discuss a business venture they are considering together. Peter takes the train home. He is never out of touch because his VPN profile extends into a virtual roaming area in his wireless operator's WAN. When he gets home, his handset selects the PBS instead of the nearest cell site just outside. His home PAN includes the PBS, most of the appliances, his TV, and his home information terminal, which is part of the *info-net* service option he purchased from his communications provider. Peter's neighbor has a VPN profile with the same communications provider, but it is quite different from Peter's VPN profile. We don't know the neighbor's name, because he's an undercover police officer. The local police have their own secure radio system throughout Albuquerque. The system is a PMR system that has its own proprietary and secure radio protocol. Peter's neighbor has the same type of mobile terminal Peter has. In fact, they both bought their terminals in the same store at the same time and had their VPNs set up on the same day. Peter has visited his neighbor's home and has seen that his neighbor is somehow linked to his PMR in a similar way Peter is linked to his employer's LAN at work. However, Peter doesn't know much more than this, as the local communications provider is well-known and highly respected for its security features and practices. Security has been an important concern for just about everyone in town because public life has not been very safe anywhere for years.

Peter gets lots of e-mail, which he usually reads on his workstation in the office, but he finds his home information center satisfactory for viewing some of the e-mail, which accumulates during the hour-long ride home. Peter is like lots of people these days: There is no distinction between work and home. Peter is waiting for a special message from his friend who is flying in to the airport, which is only 10 minutes by car from Peter's home. The e-mail will include the arrival information together with the details of a business deal the two have been working on for a few months. There's plenty of time—enough time to get the laundry started in the washing machine, including the shirt with the grape juice stain he's been wearing all day. While the washing machine is filling up with water, he works through an application on his home *info-net* terminal designed to give special instructions to his washing machine. The application eventually creates a message addressed to the washing machine, informing it of the grape juice stain, and asking that the machine to take care of it in an appropriate manner.

Since the weather outside looks a bit threatening, he turns on the TV to see the latest weather report. He knows the long-awaited e-mail has arrived, earlier then he expected, because a small envelope icon appeared in the upper left corner of the TV screen. Now, since he's in a fully converged PAN at home, he has a choice. He can read a condensed version of the message right on the TV screen, or he can read the whole message on his *info-net* terminal in the wall. Anxious to discover his friend's arrival information, he chooses the condensed form on the TV screen with a voice command, "View it!" If Peter had instead said, "Read it!" the TV would have read the condensed e-mail to him with a synthesized voice. It's better to read flight schedules than to listen to them.

Peter's friend has apparently taken an earlier flight, and has already landed, but there is no gate information in the message. A small washing machine icon flashes on the TV screen for an instant as the machine asks an information server, somewhere, for instructions on removing the grape juice stain. Seeing from the weather report on the TV that the weather looks much worse than it actually is, he starts out to his car to make his way to the airport. Peter's ever-present wireless terminal includes an optional security application, which will not allow Peter's key to work in his car or even in the front door of his home without his wireless terminal being very close by. The car and home security application has two parts. The first part brings Peter's car or his front door into a local RF network similar to the way the emergency medical application included the medical sensor his grandmother wore around her neck. The second part includes the wireless network's security features and its ability to track a user's location and other habits. If Peter correctly enters his PIN into his terminal, and all else seems normal and in order according to the network, then the network sends a command to enable the lock. Since Peter travels a great deal in his car, he has a special permit from the local government allowing him to use a hybrid electric/fuel car. Most people can only get a permit for a full-electric car, which is okay with most people, given the $33 per gallon price for fuel these days. But Peter's hybrid car is very attractive to thieves, who don't seem to worry too much about fuel prices when they steal cars. As Peter settles safely in his car, he points to his destination on the touch-sensitive map displayed on his car's information display: the local airport. The map disappears and almost immediately reappears with the current travel hazards displayed with vivid icons and real-time videos: a closed road due to an accident, two areas of violent gang activity, and a tunnel fire. The recommended route to the airport is highlighted, which Peter accepts as he drives out into the evening.

Peter's wireless terminal, which is now attached to the WAN outside his home, rings. He answers only to hear the synthesized "voice" of his washing machine complaining that it cannot turn off the cold water in the rinse cycle. The washing machine offers a selection of options: (1) come home and take care of the situation, (2) allow the washing machine to close the main water supply valve to the house, or (3) let the washing machine call the emergency plumber right away. Peter pulls off to the side of the road as he selects his security feature on the terminal. A map of his home's floor plan appears. Peter touches the camera icon in the utility room so that he can see what's going on with the washing machine. Everything seems to be in order, as no water is visible on the floor. Peter selects option "2." He will deal with the crisis himself when he gets back home.

Peter starts up the arrivals ramp at the airport just when the information display in his car illuminates with his friend's arrival gate. Peter's VPN profile includes a detailed PAN description, enabled by distributed, artificially intelligent mobility nodes, that sense Peter's location and the available wireless devices he has at the moment, and forwards appropriately formatted messages to the *best* device. Peter has been using his VPN profile for two years, during which time the network has learned a great deal about Peter's behavior, enough to where Peter's PAN description, which is created by the network itself, can make sophisticated decisions on message routing and formatting, as well as what kinds of messages (voice, data, or video) to store for later use or even discard.

Peter picks up his friend at the airport and starts on his way back home. Peter would normally expect his VPN profile to start a pot of hot water for tea at home, but the system knows the main water valve is closed for now. Peter knows this too, but still appreciates the reminder of this trivial fact on his car's information display. Peter made reservations for his friend at a local hotel more than two weeks ago, so this is a good time to go straight there, perhaps for a drink in the bar. As Peter's car nears the hotel, another message appears on the car's information display. It's from the hotel, which cannot honor the reservation because of a water problem in the local area. But, alternative arrangements have been made at another hotel not too far way. Peter doesn't know this alternate hotel, but a map was included with the message, and Peter's VPN profile calls for GPS-enabled driving instructions for any map sent to him in his car.

Peter and his friend have been in the hotel's bar only 20 minutes when Peter's handset rings again. It's the washing machine calling again. This time the washing machine's synthesized voice explains that all is well with the cold water supply after all. There was, apparently, an interruption in the main supply, not a break or some other catastrophe. The machine will, with Peter's approval, restart the rinse cycle. Peter thinks the water main interruption is probably what closed the hotel with which he made his original reservation, but he's not sure. There have been lots of strange things happening with the water supply ever since the water supply poisoning scares of 2012. The security features installed in most of the municipal water systems still cause occasional outages, but most people don't seem to mind much; just about anything is better than the random scares of the past. "Okay, do it!" he replies to the washing machine. Peter and his friend complete their schedule plans for their visit together, and Peter sends it to his scheduling calendar at the office, through his wireless terminal, just as he gets his friend checked in to the hotel. Peter arrives home a few minutes later to see that the laundry is waiting to be moved to the dryer. Peter is happy to see that the grape juice stain is gone from the shirt as he places it in the dryer.

The world seems a happy and orderly place to Peter, who issues a few commands on his home information center before turning in for the night. One of the commands authorized a monthly payment to his communications provider, the same provider that set up his VPN two years ago and maintains his PAN description. Someone told Peter once that his communications provider was not actually the network operator, but he wasn't sure what that meant. Peter's communications bill is one of the largest he pays each month. However, that's okay with Peter, for his monthly service charge includes *all* his communications, security, and entertainment needs. He also signed up for the e-commerce option six months ago, and, given the high price of fuel, it's been a real money-saver for him. The monthly bill was very small for the first year as his PAN description was maturing in the network. If we were to ask him, Peter would recall how strange things were in the first six months: The TV was calling him at work to remind him about TV shows he didn't care about, and the washing machine was leaving nonsense messages at his hotel when he was traveling away on business. Things started to get better when his communications provider assigned Peter his own private consultant. Her name is Megan. She's great. She's always available when Peter calls her; she knows every detail of the network and is always familiar with Peter's VPN profile and the current state of his PAN description; and she can make adjustments instantly. She can even respond to e-mail Peter sends her from time to time. Megan called Peter recently on a video call to tell him about a new model wireless terminal he can buy and to assure him that all the contact information and other personal files in his present terminal are all safely backed up in the network. She further assured Peter that if he decided to buy the new terminal his complete terminal personality would be loaded into it with no problems at all. This is a good thing, according to Peter, because he personally does not know anyone who knows how to reach people, or *dial* people as folks used to say, without a personal wireless terminal. People have been selecting names on various displays for years, and the networks have been finding the selected individuals, services, and companies without problems. Peter sees old movies from the 1990s and the 2000s on TV, in which people dial *phone numbers* and type in *e-mail addresses*. Megan told Peter that he had some e-mail addresses of his own, but he couldn't remember them after Megan showed them to him. Peter was a little disappointed to learn that Megan was made of about 800 polygons, and existed only in software, but the news explained some puzzling things such as her 24-hour availability. Things are great now. Peter would tell us that he doesn't mind paying the higher bills these days. The service is much better, and the sense of security is comforting. We could confirm this with Peter, but he's asleep now, and we need to see how all this came to be.

How, indeed! Less then 40 years ago, in the early 1980s, you had to know somebody or have political influence before you could own a car phone. How did we come so far? How did we come to the place where people like Peter can wander through life with safety and convenience?

Since time corrodes the lessons learned from the past, it is never a good idea to move into the future looking backward. But given the tendency of human endeavor toward folly and false starts, we will use past successful experience to separate what ought to happen from the larger universe of things that could happen. We will confine our attention to the U.S. examples from its cellular past. The reader is free to translate these into his or her own circumstances.

11.3   The precellular age

In 1976, only a few years before the launch of AMPS in the United States, the IMTS in the New York City region covered an area of roughly

50 miles radius from the center of the city, thus covering about 20 million inhabitants with RF from a bank of 12 centrally located base stations. Only 543 people subscribed to the precellular mobile phone service, with 3,100 on a waiting list even with a 50% blocking rate during the busy hour—and you couldn't even roam outside your home coverage area. There were 12,000 IMTS subscribers in the United States in 1976. The name of the technology implies that there was an earlier technology in need of improvement. Indeed, IMTS was a considerable improvement over the earlier mobile telephone service (MTS) protocol. The IMTS protocol removed the mobile operator from the call setups, relieved the mobile users of their idle channel search routines, and added a measure of privacy to the system.

In contrast with the monopoly market for fixed telephony services, wireless service has a long tradition of competition in the United States. Most markets had two operators: One was AT&T, the former telecommunications monopoly, and the other was a competing carrier of some kind. The IMTS-based mobile phone system was one of two precellular mobile phone systems in the New York City area and in general use in the United States. The competing provider used similar technologies, which were lumped together under the Radio Common Carrier (RCC) label. The situation for the RCC technology was roughly identical to that for IMTS, and will thus not be discussed in this chapter. A waiting list of 3,100 for a subscriber base of only 543 users would qualify as a successful system by most standards, but MTS was still a dispatch-based land mobile radio system with wireline interconnect, not altogether different from mobile telephony of the previous 20 years.

11.4 The first generation

IMTS could not satisfy the enormous pent-up demand for untethered telephony without a significant departure from its old land mobile origins. More than 10 years of political squabbling would pass between 1968 when the FCC proposed to allocate 70 MHz of spectrum in the 800 MHz band for a *cellular* system, and April 9, 1981, when the FCC finally approved the concept of a wireless cellular phone service. The initial permits for AMPS networks were issued in November 1982 for deployments in the 800-MHz band: 824–849 MHz mobile transmit (reverse link), and 869–894 MHz base transmit (forward link). Similar systems [e.g., NMT and total access communication system (TACS)] were licensed throughout the industrialized world. The cellular proposal suggested a system (AMPS), exclusively for AT&T, of numerous, small contiguous radio coverage areas called cells, whereby identical radio frequencies could be simultaneously reused by several low-power base stations in radio sites placed relatively close to the ground rather than high atop a structure in the center of a huge coverage area. Each of the cell sites would be miniature copies of a more traditional radio site, but each would be linked to a central trunking controller, also called a mobile services switching center (MSC). Different cells would carry different conversations. The number of mobile phones could be substantially increased without a corresponding increase in the number of required radio frequencies. Also, replacing the central radio site with hundreds of low-powered cell sites, which adds a spatial dimension to the trunking protocol, allowed the mobile station's transmitter power to be lowered to a point where the automobile's luggage compartment did not have to be reserved exclusively for the phone [1]. The portable terminals that quickly appeared changed mobile telephony from an alternative service in places where wireline service was not available to a service that was complementary to wireline: A salesperson could be reached on the road with wireless service just as easily as he or she could be reached in the office on his or her wireline phone.

The first-generation cellular systems were unqualified successes everywhere they appeared. Figure 11.1 shows the dramatic results for the AMPS system in the United States. The cellular industry regularly underestimated its own growth throughout the 1980s.

Even with the stunning success, all was not well in the first-generation analog networks. There were four salient problems:

1. FM radio limited the capacity of the systems, and it limited the ability to bring emerging network features to subscribers' handsets.

2. The commoditization of FM cut the revenue reward to those who developed the technology.

3. The frequency reuse scheme did not work as planned; it was too expensive and difficult to manage C/I as cells approached radii of about 2 km.

4. The stifling business environment kept airtime prices high and limited growth.

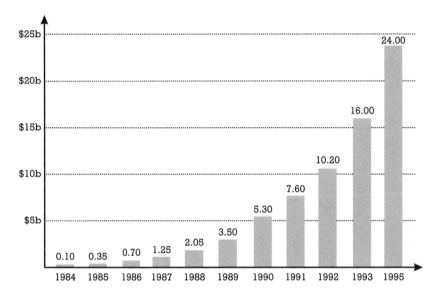

Figure 11.1 Growth in cellular subscriptions. (The vertical axis represents annual expenditure on cellular communications in the United States.) (*Source:* CTIA.)

It's easy to make a strong case for second-generation digital technologies that offer tools for real competition if you view the second-generation technologies as adaptations of first-generation analog systems to a competitive business environment. You can also declare the second generation as a logical extension of the first generation that tracked the parallel evolution in the fixed networks.

11.5   The second generation

Six operators in a market deploying different types of digital cellular networks brought two kinds of real competition to the industry. One kind was the ability to differentiate one network from another with new services. All three of the digital cellular standards (GSM, TDMA, and CDMA) provided at least the possibility of new teleservices (SMS is an example), bearer services (circuit-switched data is a salient example), and supplementary services (caller ID and call forwarding are examples). Digital radio provided a means to bring the fixed, generic digital network[2] out to the handset so that wireless service continued to resemble the telephone service the user enjoyed at home or in the office. The second type of new competition was the sharp increase in the number of operators. Without a huge increase in demand, all the new operators would not survive. This second factor turned the cellular landscape from an orderly one of marginal competencies into a war zone in which airtime prices fell to wireline levels in many markets. Cellular was now a substitute for wireline service, particularly in places where the waiting time for new wireline service was months or years.

Second-generation digital cellular can be viewed as an evolution of first-generation analog technology that allowed wireless service to track the advances in the fixed networks. The fixed operators had done a remarkable job of constructing VPNs for subscribers: voice mail notifications, call forwarding, and bundled services. Digital radio increased the efficiency with which a wireless VPN could be constructed for a user, and offered enough planning tools and service options to allow operators to react to new competition.

By 1999, everyone who wanted a phone had one in many markets. Some households had multiple subscriptions, and some markets (e.g., Finland) had wireless penetration exceeding wireline penetration. New and unpredicted wireless applications appeared. But the good times appeared to be coming to an end. The average revenue per user (ARPU) was falling as predominantly casual or emergency-only subscribers were encouraged to subscribe at lower rates. Clever schemes, which were all enabled by the generic digital network, expanded the fading market even further. Prepaid, for example, brought even better account control for some users, and offered the possibility of wireless service to those who used to be excluded because of credit problems.

Competitors would have to depend on something other than market growth for new subscribers. What is left? One source is the competitor's subscribers, which could include those belonging to the local wireline operator. Another more attractive source comes from expanding the market with new data services. This second alternative has the added advantage of limiting churn. An operator can raise barriers to churn such as (1) offering unique and attractive service offerings not available

---

2. A generic network is one in which the user's data and signaling are represented by discrete, digital signals everywhere. This means that the value of a generic network can be increased merely by breaking one of the interfaces between two nodes and inserting some kind of bit manipulator. It is relatively easy to enhance the functions in a generic network. Digital radio extends the generic network out to the user's terminal or handset.

through the competitor and (2) erecting all kinds of username and password hassles, which could discourage churn. The industry considered the new data service alternatives as a logical extension of what had already been going on in the fixed networks since the early 1990s: the Internet. Many watched the dramatic effects the Internet was having on business and society and noted that the enormous changes were enabled by relatively simple and low-cost devices. It is impossible to overstate the effect the modest little modem has had on culture, politics, and businesses. What effect would fiber to the home bring? Nobody wanted to wait to find out. Nobody wanted to be excluded from the Internet.

11.6 The third generation

The bandwidth in the fixed networks is already so large that it costs more to bill a long-distance call than to deliver it: Zero-cost long distance is already here. The third-generation systems extend their second-generation predecessors just like the second generation extended the first-generation analog technologies: Each let wireless service track the bandwidth increases and other changes in the wireline networks. However, the first- and second-generation technologies were all responses to obvious market demand. The first generation broke the severe capacity constraints of the precellular technologies. The second-generation digital technologies were elaborate rescue missions for analog systems, which were eventually threatened by the evolving tastes and preferences of fixed-network subscribers and the appearance of real competition in the place of the original monopolies and duopolies. The third-generation initiatives are a risky departure from this old tradition of incremental change driven by obvious market demands. It is an internal industry initiative, which seeks to create new demand—demand for new services to be sold to old customers, and demand for new services attractive to new wireless users. There are four stages to mobile wireless evolution, as shown in Figure 11.2:

1. The *alternative* stage is the market entry stage. It is the stage where wireless is only considered when wireline is not available: Cellular-based WLLs in rural areas are a modern example.

2. The *complementary* stage is the optimization stage. It is the stage where falling prices force highly optimized systems that are

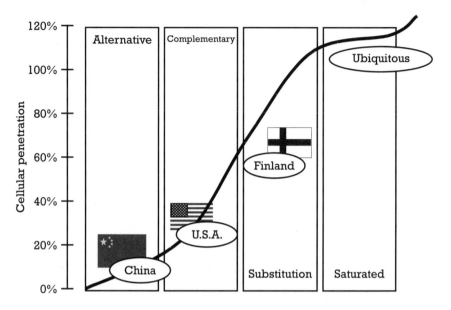

Figure 11.2   Four stages in mobile wireless evolution.

attractive to users who appreciate wireless only some of the time (e.g., while commuting or calling on customers).

3. The *substitution* stage displaces wireline service. Some of the subscribers in mature second-generation systems do not have wireline service all.

4. The *saturation* stage is often considered the last and most mature stage. It is the stage where wireless terminals are ubiquitous for voice and data across multiple devices.

A substantial portion of the wireless economy is getting mired down in the substitution stage. They are stuck in this phase because the technology can no longer emulate the current wireline services, notably access to the Internet. The next stage, the saturation stage, is the one where mobile telephony is ubiquitous, which implies that it extends to data, video, and multimedia applications that can sustain growth for as long as new applications can attract new revenue. Third-generation systems seek to meet three goals:

1. Bring packet-switched services—the Internet—directly to the subscriber;

2. Provide the channel capacity on the radio interface to accommodate multimedia;

3. Preserve the investments in second-generation networks.

The first is satisfied with the installation of packet routers and associated databases and the interfaces between them, which can route packets to their correct destinations. The second goal is accomplished with the addition of an adaptable radio interface, WCDMA for the GSM extension, which provides mechanisms for an enormous variety of multimedia, most of which don't even exist yet. The third goal is met with some kind of brute force mechanism specific to each radio technology. The tendency is to use the legacy radio technique as one of the 3G operating modes. 3G implies new kinds of business plans that recognize an overlap between those who use the Internet and a mobile phone. The successful plans will be those that bring price-inelastic revenue through tailored content access on a per-user basis: the so-called wireless Web. Early notions of the wireless Web were primitive ideas that proposed uninterrupted access to the Internet. Smart phones, feature phones, and standardized applications environments like WAP will eventually make users comfortable with the wireless data interface. Real success will come from entrepreneurs who recognize the unique aspects of wireless access, something that distinguishes wireless from fixed access: the ability of a mobile network to deliver the Web to a mobile user at exactly the right time and location. To the extent that the Web's content can be tailored to the user with a wireless personality, the business will be even better. As was the case in previous generations, 3G has its own enabling technologies:

• New source coding schemes for the nonspeech applications;

• Turbo coding to provide the low error rates required by the new source coding routines;

• Adaptive air interfaces;

• Packet radio;

• QoS (bandwidth management protocols and management systems).

Web-savvy users are surprisingly sensitive to network limitations. Early WAP experience in Germany, for example, shows that though users do not mind dialing an access number for their WAP-based services, they are annoyed with having to pay for the connection while information is not flowing between the network and the handset. They seem to sense the difference between a dedicated connection and their nonvoice data. Here we have evidence of the possible success of packet radio (e.g., GPRS). The WAP supports low-tier applications featuring point-to-point, discrete transactions. These first tier applications (news, weather, stock quotes) will drive confidence and familiarity with the wireless Web and will become a universal set of applications in mature markets (billing inquiries, network outage notifications, traffic and weather alerts, and handset software update notifications), but they will not bring sustainable revenue. Higher-tier applications will bring the profits. In summary, WAP applications can be categorized as follows.

‣ First-tier applications are basic point-to-point services (weather, sports scores, stock quotes), which work well with a common interface that adjusts the content to the display on the handset.

‣ Second-tier applications provide PIM/PDA synchronization (address books, calendars, to-do lists), which is a sophisticated and secure form of the first-tier services.

‣ Third-tier applications include personal information (bank accounts, travel reservations). These kinds of applications will thrive on application-specific terminals, or general use terminals with proprietary software.

‣ Fourth-tier applications are transaction-based and fee-based services such as location-based wireless e-commerce.

Where will the higher-tier applications come from? They will not come from network operators who will initially fail to see an inevitable separation of infrastructure ownership and management from service delivery and marketing. The wireless access service provider (WASP) will become the subscriber's interface to the network operator, and the best WASPs will include customer support and billing in the applications realm. There will be plenty to keep the operators busy in 3G. With packet radio, for example, comes applications with chunky bandwidth demands, which will have to temporarily reallocate capacity away from lightly loaded cells to those with high bandwidth demands. Since capacity is power-limited in 3G networks, wireless bandwidth management (wireless QoS) could allow cell breathing.[3] 3G is all about variety and non-standard terminals, each with its own radio environment, software architectures, and unique needs. The network operator will have to give the WASPs the ability to manipulate transport channels and wireless QoS. When the diverging needs of the content-centric Web can be married with the device-centric wireless applications, 3G will be successful. To the extent we can find new services about which users care enough to pay, and deliver them efficiently, we can limit the risks in 3G. The approach should mimic past evolutions. First, concentrate on the customer, then manage the network on behalf of the applications. Mobile network operators need to surrender their subscribers to WASPs. They need to see the user's needs in the WASP's demands, and keep the customer's attention on the WASP and its applications. The wireless operator needs to focus on the unique value the mobile network holds for the WASP.

1. The mobile network has the ability to determine a user's location. The subscriber's location is valuable to the WASP.

2. The better a network operator becomes at providing a QoS-configurable pipe for the WASP's applications, the more valuable the wireless infrastructure is to the WASP.

Finding the user's location enables location-based services: traffic reports, hotel and service information, maps, driving instructions, toll road payments, emergency services, and location-based advertising. User locations are an asset that can be sold by the network operator to its content providers or a third party, but there are problems with determining location. First, current methods are not very precise. The cell of origin (COO) method can have an uncertainty as much as 30 km. Second, many location-based services require heading information in addition to location. It doesn't make much sense to offer driving instructions to the

---

3. In CDMA-based radio networks, as the cell loading increases so does the interference. This tends to result in those on the edge of the cells experiencing a lower signal-to-interference ratio and in some cases, not being able to communicate. In this case, the cell is said to "breathe" as it effectively expands and contracts depending on loading. By more tightly controlling the amount of interference generated through the mechanisms discussed above, either breathing can be reduced, or compensated for.

nearest gas station if that station is off the freeway exit that the subscriber can see in his or her rearview mirror. It is difficult to determine heading without frequent updates.

Without QoS-configurable pipes for applications, 3G networks could fail even if they are not content limited and have hundreds of products for users. Networks can be content limited by, for instance, the WAP interface (which leaves the application in the first-tier), or they can be too dependent on third parties such as terminal makers and marginal content providers. They can also be content-limited because the operators try to provide the content themselves, leaving the subscribers with few choices and not enough reasons to remain loyal customers. Operators will probably succeed as wireless information pipe suppliers and leave content, both the transactional (e-commerce) and the advertising types, to others who know how to do it. Network operators will have to restrict their subscriber personalization work to allowing third parties to customize the wireless pipe to a specific user, application, or device.

Will new and larger networks be more valuable by themselves? Metcalf's Law[4] says that the value of a network increases with the square of the number of its nodes. If a network with 10 nodes is found to be worth $100, then expanding that network to 100 nodes will increase the network's value to $10,000. Aside from the troublesome radio interface, mobile networks are different from fixed networks in two other ways. First, most of the nodes in a mobile wireless network are the user's terminals, which are completely content-limited: They are only used if there is a reason to use them. If there is no compelling reason to use a mobile terminal, then the user is free to go to a terminal in a fixed network, which will remain an attractive alternative as the bandwidth in the fixed networks increases. Second, the ability of the network to follow a subscriber's movements among its nodes is more valuable than the nodes themselves. Look at the way most of us use our fixed line telecommunications services today. In the fixed realm people reach us by dialing a directory number, which is mapped to some kind of device at a fixed location. We actually have to complete some rather sophisticated management tasks to get our information to the proper location: We have to consider geographic location and find the correct device for the traffic we are trying to move before we can select the correct directory number.

---

4. Bob Metcalf of Xerox is credited with the invention of Ethernet, which is a CSMA/CD protocol implemented in layers 1 and 2. It is described in IEEE 802.3.

Even in the relatively simple case of teleservices, we need to make sure we ring the phone when we want to talk to someone, and send our fax data to the fax machine when we don't want to talk on the phone. We either have to make sure we have the appropriate directory number for the intended device we are trying to reach or hope there is some kind of clever contraption on the terminating end that can connect the proper device to the socket to which we are finally connected. Compare these exercises we do all the time with getting our e-mail while traveling. Once we figure out the proper connector and cabling to use between the laptop and the connector we find under the desk in the hotel room, go through all the dialing experiments, dial-up network settings, and modem settings, the Internet seems to find us with little trouble.[5] The tedium is largely confined to establishing a connection to the local PSTN. The hard part—finding any one of us among the millions who travel everyday—usually works well. What would happen if we used e-mail addresses for everything? It would not matter where we were or what device was handy: PC, phone, computer kiosk—the Internet knows which device to "ring." The lingering confusion and difficulties come from the legacy PSTN with its location-based dialing and analog spans out at the edges of the network. Addresses become valuable assets when they can be counted on to find people. AOL, Yahoo!, and MS Hotmail have the kinds of addresses that find people easily, and these addresses can be sold.

Eventually, users may wonder why their e-mail can't follow them seamlessly to other devices attached to other kinds of networks, both fixed and wireless, which is the users' perspective of *any content, anywhere, any device*, the ultimate in mobility management. If users start wondering when seamless application will appear, then we can declare the 3G

5. This writer travels frequently to many countries and has become weary and impatient with the tedious and expensive (both in terms of room charges for each modem setting experiment, and in terms of personal productivity) in-room telecommunications experiments that must be completed upon check-in. A satisfactory alternative seems to be a GSM phone with the appropriate cable and terminal adapter. This scheme works well if you do two things. First, carefully select a multiband GSM handset, or confine yourself to a *family* of handsets (900- and 1,900-MHz types) from the same manufacturer that use identical data cables, terminal adapters, and configurations. Second, you complete all of your connection exercises, at home on your home GSM network, only once. Then, when it's time to contemplate retrieving your e-mail in some far off land, you (1) check with your home GSM operator for a roaming agreement with an operator in the country you plan to visit, and (2) make sure your hotel room has adequate RF coverage before you accept the room.

adventure a success. Every adult—and most kids—will have one or two wireless terminals running a dozen applications. Now what? We have to look beyond 3G to keep everyone in the wireless industry employed and the users happily looking for even more. We have to move each subscriber to hundreds of terminals running thousands of application, from point-to-point transactions to a constant blizzard of multipoint-to-multipoint services.

11.7   The post-3G world

Imagine what could happen if the 3G world extended the Internet's ability to find people to any of their devices and applications. This is an important exercise in converged networks, in which applications take specific advantage of the mobile network's ability to determine location. General Motor's OnStar™ system is a primitive example. What happens if we go beyond the WAN, the cellular networks that support OnStar™ for example, and include LANs and PANs? Peter's experiences in the beginning of this chapter include many examples. The post-3G world will be one in which subscribers are *always* attached to the network, perhaps not always exchanging data with the network, but permanently attached. It will be a world where the ubiquity of wireless terminals will have extended far beyond the user who pays the bills. It will include users' most humble possessions: their cars, their washing machines and refrigerators, and even the lock on their front doors. If subscribers are not attached themselves, then they will be attached through one or many of the agents (appliances) in their PANs. Subscribers will be virtually present and permanently known to the network: Users will be *e-present*, and operators will strive to bring their customers to a permanent state of *e-presence*.

Figure 11.3 is a review of commercial mobile network evolution. A clear progression is evident only if we confine our attention to the user's needs and how the industry met them. The transition from precellular to the first cellular systems solved obvious and pressing capacity problems. The next transition into digital systems kept the party going by attracting even more users with additional capacity, lower prices, and personal services that resembled the new ones, which were constantly appearing in the fixed networks.

The movement from the second generation to the third is a new experience for the industry. It is one that does not directly respond

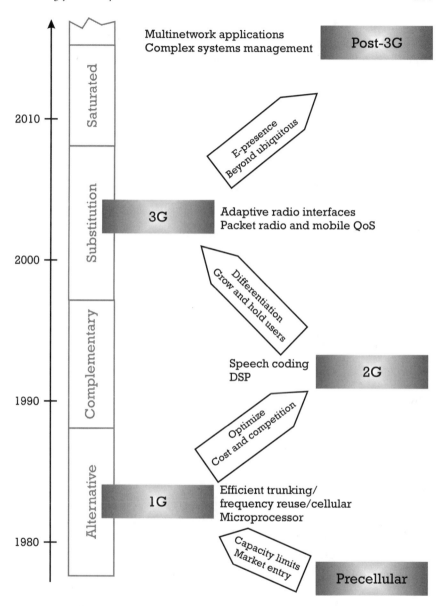

Figure 11.3 Mobile wireless evolution.

to external demands from customers. This time it is looking for ways to expand its own markets and create new demand on its own. It tries to deliberately drive the business into a saturated market without an obvious demand from users. Vague initiatives and business plans that draw on packet-switched services and multimedia are touted and celebrated. Viable business plans and specific applications that would be attractive to mobile subscribers are absent. If the 3G adventure fails, then *wireless* could become a synonym for *jobless*. Any path beyond 3G will have to abandon the logical extensions, one after another, that characterized the earlier transitions, and return to the customer's needs.

11.8 Constraints on the mobile wireless future

Four factors will threaten the likelihood that the universal system illustrated with Peter's wireless life at the start of this chapter will ever appear:

1. Standards;

2. Security;

3. Falling revenue and limited capital;

4. Lack of vision and talent.

11.8.1 Standards

Today's 3G Partnership Program (3GPP) and 3GPP2 committees can be viewed as catalysts that accelerate the slow chemistry of the standards formulation process. They also impose a welcome international character to the work. Standards traditionally expand markets and lower costs by assuring equipment from different vendors will operate with each other. The synchronous digital hierarchy (SDH)/Sonet multiplexing scheme is a recent triumph in standardization. However, standards can also inhibit innovation, which thrives in proprietary spaces. The WAP is an example of a confining standard; it still assumes the one-terminal-for-everything model of traditional cellular practice. Wireless multimedia is all about many different kinds of application-specific terminals and appliances. The standardization process will have to change from one that *assures compatibility* to a process that *enables incompatibility*. Emerging standards need to foster lots of variety in applications and appliances, and particularly favor proprietary innovations. They need to recognize, accept, and enhance the generic digital networks. Operators will need to deploy systems that can accommodate just about anything. The best standards will be the ones that can accept the most proprietary innovation and give network operators the tools to let content providers manage the available bandwidth themselves. This implies a self-healing and self-optimizing network.

11.8.2 Security

Security will be both the enabler and inhibitor of the post-3G world. Peter lives in a world of stolen cars, tunnel fires, violent gang activity, and constant terrorist threats to the municipal water supply. He also lives in a world where many chronic diseases have been conquered through automatic and assured attendance by qualified medical personnel. The same theories and mentality that have poorly trained and low-paid people protecting scores of multimillion dollar aircraft in today's airports by asking inane questions at security checkpoints also suggest that communications channels can continue to protect traffic only if the security mechanisms remain shrouded in secrecy. New and useful content attractive to those who will live in a dangerous world can only thrive in secure networks. Will the telecommunications industry, the regulators, and governments finally understand that the best security is assured for everyone with open standards and processes, which are freely shared across borders?

11.8.3 Falling revenue and limited capital

The myth of limited spectrum for mobile telephony inflated the value of mobile wireless networks until advanced digital radio techniques increased capacity beyond the artificial limits set by the regulators. Third-generation systems provide enough mechanisms for operators to distinguish themselves from each of their competitors by freeing them from the trap of competing entirely with voice telephony. 3G will make the competitive environment that emerged with the second-generation networks even more intense. The operators will have to attract and retain subscribers with unique content, which can be enabled with an efficient network in terms of coverage, bandwidth management (QoS), and customer service. The content, not the new terminals, will attract the new revenue. Standards that confine content, like the WAP, will lose importance except for common housekeeping services. The value of the networks will not reside in the licenses or the potential number of subscribers, but in their ability to earn revenue. The sources of capital will be impatient with those who cannot show early and consistent revenue growth. The critical part to assuring revenue will be the content that can attract and retain users. All but the best operators will fail because their systems will remain content limited because of a lack of talent and vision.

11.8.4 Lack of vision and talent

The best way to have a good idea is to have lots of ideas. Some of those have to be good ideas, and good ones are going to be in very short supply when wireless economies move into the ubiquitous state. To the extent that 3G is successful, wireless terminals of many kinds will be everywhere. Penetration will be much more than 100% because each subscriber will have more than one wireless appliance. But this will put the whole industry into the upper right-hand corner of Figure 11.3: a saturated market with no place to grow. The industry will have to turn its attention to creating a steady stream of new and fresh content; new applications that beget new appliances and services. How long can you throw new content into a network and still come up with fresh ideas that will attract new revenue—in Web time—when you are in a saturated market? One way to keep the doors open is to find ways to take the WAN's content off of other networks: LANs, PANs, and PMRs. There are plenty of examples in Peter's wireless life. However, this demands that the practitioners abandon almost everything with which they are comfortable—a way of life attractive only to bold visionaries who are comfortable with permanent uncertainty and lots of risks. It also suggests some new enabling technologies.

11.9 Conclusion

What will be the enabling technologies for the future-generation cellular technologies? We can answer this question by reviewing the enablers for the earlier evolutions: better trunking efficiency, frequency reuse, advanced source, and channel coding. These all directly served the user's needs. The future will find the user in many kinds of networks at different times. Old concepts such as circuit switching, attaching to a network, and discrete applications and content will not apply to the post-3G subscriber. Even such concepts as a *subscriber* or a *user* may have to be replaced with *participator* or a *grazer*. Consider four enablers:

1. Morphing an application's content to a particular device or appliance;

2. Supervising multiple connections—end-to-end and multipoint-to-multipoint—across different types of networks while still managing the bandwidth across all networks for a single application, or a suite of applications;

3. Aggregating small applications into mega-applications—automatically;

4. Managing huge, adaptive, complex systems.

The good news is that some of the companion networks will already exist when the 3G effort finally has to extend outside the WANs. The PMRs will have already been established for years and PANs will emerge at the same time. A user has many different personalities and specific needs in his or her virtual PAN; these needs change quickly with time, location, and circumstances. The aggregating of many different kinds of networks into hundreds of millions of virtual PANs echoes the transformation of the simple precellular systems into the first cellular networks. Just as the cellular concept with its frequency reuse schemes was called the *breeder reactor* of frequency channels, the concept of many tiny PANs, one for each *participator*, could provide enough space for new applications well into the post-Shannon era.

Reference

[1] Webb, W., *Understanding Cellular Radio*, Norwood, MA: Artech House, 1999.

CHAPTER

12

Contents

When my kids are in their twenties

Siegmund Redl

12.1 Introduction

Writing about my personal view of "the shape of things to come" in wireless communications proved to be an exciting exercise! However, making predictions 20 years ahead is a daring exercise as well. We need only to look at the way the whole world communicates and deals with information today—who could have imagined that 20 years ago?

Wireless communications, and in particular personal communications, has developed rapidly over the past two decades, with all its measurable parameters making steep growth curves. Presenting them mathematically or displaying them graphically would require the use of exponential terms.

Let us pick just a few of these parameters:

> *Subscriber numbers* regularly exceeded predictions and are now comparable to fixed line subscriber levels in some countries.

> *User equipment* (phones, terminals, etc.) made stunning progress in size and performance, starting as bulky in-car installations or portable handbag-sized devices with limited battery life and moving toward small pocketable gadgets with up to 700 hours of standby and features such as efficient data transmissions and Internet access without wires.

> *Technological progress,* which was partially fed by the demand to further shrink devices and to squeeze in more functionality and performance, enabled all those developments. Silicon technology—on the basis of the technological value chain—has evolved tremendously and continues to do so. This is a manifestation of Moore's Law, proposed by Gordon Moore, cofounder of Intel, in 1965 (see Section 4.2). Also, the progress in other engineering domains like system engineering, software engineering, radio engineering, display/user interface technology development, as well as in mechanical and manufacturing engineering, is part of this process that continues to feed on itself. The industry continues to produce smaller, cheaper, and yet more powerful wireless products in shorter development and innovation cycles.

In parallel to the rapid technical evolution that enabled and attracted wireless users, there was also a social evolution. The distribution of and the access to information have been accelerated through the Internet explosion of the 1990s. Users in both domains—the cellular networks and the Internet—initially were exotic creatures, phone buffs, and computer freaks. Today, the use of cellular phones and of the Internet is common among all professions, ages, and interests.

With new generations of users growing up, the average individual in our societies is more and more used to dealing with technical equipment such as a VCR, camcorder, CD player and recorder, DVD player, PC, and digital camera. He or she is increasingly Internet-literate (that is, able to access the Internet, communicate through it, and retrieve and present information through it). Social groups that were excluded from the use of modern communications technology and the Internet society increasingly find access to it—for example, through special educational programs (schools, government campaigns) and as usage becomes more affordable with the economy of scale of true mass media.

12.2 A look back

Before we attempt to take a look into the future of wireless, it is important to also look into the past and present, which helps us to understand the dynamics and parameters of spreading wireless technology and wireless-enabled applications around the world. There are differences we can already see today in the use of mobile communications services, mostly visible through the regional particularities in usage, experiences, and expectations, and through different generations, again with different expectations and different attitudes.

When I started to work in this exciting industry, in 1990, the proliferation of wireless communications—then exclusively using analog technology—was limited to professional use in dedicated PMR networks and in first-generation cellular networks. Use of cellular phones was limited to mobile professionals and groups of wealthy individuals able to afford the premium cost of mobile phone services. Then came the so-called second-generation digital technologies, namely, GSM, CDMA, TDMA, and other open industry standards, and with them came a vision of communication anywhere, anytime, and for anyone.

Up until 2000, the use of second-generation (digital) cellular networks for data services had not become a significant factor in either commercial or usage terms. The only really successful data service was the SMS within GSM networks, and this was only really used by teenagers. Sending short messages (SMS) instead of calling has become a fashion that led operators to offer certain billing/tariff models for these target groups. Some individuals spend more money on SMS than on voice calls, even though a single short message is cheaper than a quick voice call.

There are many reasons for the limited use of circuit-switched and packet-switched data services, a thorough analysis of which would go beyond the scope of this work. One of the main factors is the generations of users. Most initial users were individuals who were just taking advantage of the fact that they could initiate and receive phone calls from just about any place away from home. In some countries and some cases, the mobile phone was used in place of the fixed line phone at home. All this was seasoned with the introduction of so-called supplementary services known from digital circuit-switched networks (call divert, etc.) and other features. The general user group did not (yet) have any need for or any understanding of data services. This, in combination with the quick acceptance of general mobile phone voice services, led the operators to focus on extending network capacity and the improvement of quality for voice services. The industry attention to and the hype around the introduction of enhanced voice coding technology and other improvements (e.g., echo-free connections) in this respect was higher than the introduction of mobile data services, even though some equipment (PCMCIA cards, data and fax transmission software and hardware) existed. Another factor concerning low data usage is the limited data speeds available in second-generation mobile networks (typically 9.6 Kbps). Although the data/Internet services in Japan have proven that attractive services and applications can be developed and offered even at low speeds, comparable services, in scope and quality, have not been successfully deployed or offered at present in Western countries. The discussion arising in 1999 and 2000 around the fate of the wireless application protocol (WAP) has illustrated some of the issues. The high expectations that were set for a widespread standardized access to content and personalized information have been disappointed by immature early products. The image of a scaled down, poor wireless substitute for Internet access does not help either. But, rather than scaling down readily available Internet services and applications, which typically run on much higher data rates than available through 2G and 2.5G technology, the key to success for mobile use is to develop these mobile services for this particular environment from the start.

12.3   The transition from 2G to 3G

The technical evolution from second- to third-generation wireless technologies and the evolution of the Internet are converging for a new age of communications.

Despite the limited data rates, some data-based applications, including limited access to Internet, are being trialed in advanced 2.5G networks, with mixed success. These may spur a wave of new data applications, although it is difficult to tell what these applications will be and which will thrive and survive. Many things will need to be tried out; some will succeed, and some will not. Who would have predicted the tremendous success of daily horoscope or comics downloads through iMode in Japan? User terminals, the good old mobile phones, are already becoming personal digital assistants, with comprehensive phone books, schedule/calendar, and data storage facilities. Increased usability comes through better technologies in the domain of man-machine interface (MMI). Improved display technologies and input technologies like voice recognition/voice dialing and overall better ergonomic designs make the mobile phone the personalized companion of more and more users worldwide. Access to certain content, again personalized through certain profiles (stock portfolio information, news/sports categories, and horoscopes), makes the use of mobile data services, in addition to the enhanced voice services, more and more popular.

Increasingly, new generations of users are becoming accustomed to the mobile delivery and presentation of such information. They are also accustomed to ways of making good use of these new services. Note that the term "generation" in this context is related to technical innovation and ways of information consumption, rather than a 20- or 30-year period of family generations.

Related technologies that are delivering data services over existing networks (discussed in more detail in Chapter 3) include WAP, packet-switched access to wireless networks such as iMode (NTT DoCoMo wireless data/Internet service, in Japan), GPRS in evolving GSM networks, and IPs over CDMA networks (e.g., EZ-Access services, also in Japan). These technologies, still limited in data transmission bandwidth to below or just at the present fixed line experience, are showing the way and preparing the markets for the higher bandwidth applications to be offered by next-generation technologies.

Many simple applications, however, do not really need the extra bandwidth. They include tracking stock values, weather, news, traffic information, even simple banking transactions or location-based and simple navigation services with alphanumerical instructions (versus graphical)—and we should not forget horoscopes!

The wireless industry and related content providers are learning that it is not enough to trim down existing Internet content format for use on low-rate wireless devices such as GPRS. Again, such content and applications need to be designed for this kind of use from the start. The image of mobile Internet being the poor relation of fixed-line (fast) Internet access would be counterproductive to its widespread adoption.

Other applications, like the timely transmission of images and maps, video clips and larger data files, are still outside the scope of the so-called 2.5 generation of mobile technologies, simply due to the lack of speed.

Some applications, though, that do not require a real-time/short-time delivery, do have a potential for success even when used at low data speeds.

Packet switching enables, and requires, different billing structures (by the byte versus by the minute), which need be introduced as well. Of course, these require well-balanced tariffs that are acceptable to the user, in particular for transactions with different quality of service levels. The billing method alone will leave a lot of room for creativity among competing service providers. Those billing models will make their way into third-generation services and will be refined further. By the way, service providers, as established in second-generation networks, do not necessarily need to own their own cellular network. By focusing on their respective strengths in providing access, billing, portal and content, and services, the players in the industry will continue to open up new business opportunities for the benefit of the user. The experiences in this transition phase of generation 2.5 of the user and the wireless industry will be important for the adoption and success of the next generation.

12.4   The next generation at the doorstep

In another attempt at harmonized and open standards, the mobile communications industry continues to work hard to make use of existing and new spectrum more efficiently, to offer higher bandwidth for data services, and to enable new applications. Mobile multimedia features and a new mobile Internet experience are promised through third-generation technologies and systems that are going to be introduced over the coming years.

Large investments are necessary to develop the technologies and make them work in a commercial consumer environment. The investments just for the plain access to radio spectrum (spectrum auctions) are being justified by the bold expectation that, indeed, there will be a high level of consumer acceptance for the new services offered. Will the math work here? As a precedent, the success of second-generation mobile networks has exceeded expectations, year after year. But the access to spectrum was, in general, free of charge to these operators. Having to pay for spectrum will clearly extend break-even periods for 3G investors, as compared to 2G.

As the industry continues to feed on its success, consumer expectations are growing, as are basic requirements for the need for service quality and falling costs. Even though there is not much room for experiments and mistakes, certain services will fail; others will be successes, unexpected at the time.

Similarly, an abundance of different user terminal types will emerge: small ones, big ones, thin ones, thick ones, ones with a built-in camera, ones with a color display, some looking more like palmtop computers, others like conventional mobile phones—others won't be visible as wireless connected devices at all. Creativity of manufacturers, driven by customer acceptance or disapproval, will be challenged more than ever before. The simple 6 × 3 keypad and the simple display mobile phone we know today will may disappear, albeit slowly. Also, new devices will be more customized by software feature installations, in the same way that applications can be added or removed from desktop and palmtop computers. Therefore, the times when the customer would only choose the color of the device and the language of user interface will be over.

Still, the one common backbone network holding all these applications together will not be a proprietary wireless infrastructure anymore. It will be the Internet, which also will continue to evolve and develop, through new features and new protocol versions that enhance its usability for mobile access. Even voice over IP (VoIP)—today not seen as a viable option for mobile voice services—will be made possible in the mid-term future. A complete replacement of real-time circuit-switched connections for all applications through IP and packet switching, even at high data rates, is uncertain today but may be possible in decades to come.

Still more radio spectrum will need to be made available in order to cope with the bandwidth demands posed by the increasing number of users and the increasing usage. However, only a convergence and thorough integration of mobile communications and the Internet makes all these future scenarios possible. Mobile Internet usage will by far surpass stationary Internet access as we see it today: through desktop PCs and squeaking modems or PCs connected to corporate servers and routers.

12.5 New technologies, features, and applications

Adding to the features and technologies mentioned before, over the next 20 years we will see a further integration of features and performance into ever fewer and smaller devices. We will also see new materials being used for battery and mechanical designs that will prolong battery life and enable different form factors, new input and output, as well as new (short-range) connection technologies. Worth mentioning here are voice recognition and voice synthesis, which will enhance the manner in which we interact with the phone, virtual displays, and BlueTooth (see Section 3.2.5). BlueTooth is a wireless short-range radio technology that enables a whole new range of applications around mobile and stationary devices, replacing cables and other interfaces such as (very) short-range infrared. The open standard and open intellectual property attributes of BlueTooth will enable it to be implemented into many electronic devices, with device-to-device communications enhancing the functionality of basic products such as refrigerators.

Yet another striking technology that will make it into these wireless consumer devices is position location. There are a number of ways to achieve position results, and each has its own distinctiveness in terms of functioning in different environments (indoors/outdoors), accuracy, and cost. They are summarized in Table 12.1.

Table 12.1
Different Types of Positioning Systems

Positioning Systems	Advantages	Disadvantages
Cell of origin: Information on the cell in which the subscriber is located	Can be achieved with no, or very minor, modifications to existing systems	Somewhat inaccurate, especially in large cells covering hundreds of square kilometers
Angle of arrival: Directional antennas at two or more cell sites determine the angle of the arriving signal and triangulate	No changes are required to the mobiles, and accuracy can be to as close as 100m	Directional antennas are required at the base stations and in dense areas the reflections may result in inaccuracy
Time difference of arrival: Synchronized base stations measure the time of arrival of a signal from a mobile and triangulate	Only software changes are required at the base stations (assuming they are already synchronized) and no changes at the mobile	At least three base stations must be able to simultaneously receive signals from mobiles; can be inaccurate where "strip" coverage is provided (e.g., along remote roads)
GPS: The mobile has a GPS receiver that determines location; it can be assisted by the network providing it information on currently available satellites	Extremely accurate, to 10m or so; limited network modifications are required	All mobiles need to be modified and GPS receivers add cost and size; does not work well inside buildings

The services and applications made possible through position location are abundant. The following are areas in which location-based services will be realized over the next five years:

▸ Security/safety/emergency-related services (like the U.S. FCC mandate for emergency phone 911 services or 511 service for traffic and travel information);

▸ Navigation services for in-car or pedestrian outdoor usage;

▸ Location-based billing and push of information and advertising;

▸ Tracking of vehicles, equipment, goods, and people.

Regarding the last item, the issue of how and when position information is being made available to whom and under which (legal) circumstances will need to be regulated and technically resolved in a way that complies with each individual's rights for privacy.

The integration of wireless devices and technologies that are already in use for consumer applications will further stimulate the abundance of new terminal devices. Moving Picture Experts Group Audio Layer 3 (MP3) audio players are already nowadays integrated with mobile phones. Applications range from simple replay to Karaoke-phones, which not only play music but also display text, as seen in Japanese devices. Other features will follow, such as audio streaming (music and spoken text replay after reception and storage in terminal) through common voice coding technologies that allow the transmission of compressed voice streams, and video streaming through integrated Moving Picture Experts Group Standard 4 (MPEG-4) decoders and through color displays. Similarly, data storage capabilities and optical sensors as required in video cameras will be integrated with cellular phones and will exploit synergy effects. Video phones, so far not widely accepted in fixed line networks, are another domain in which next generations of wireless systems can provide adequate transport mechanisms—in this case initially through fast circuit-switched connections. It remains to be seen whether this kind of application finds better acceptance in new generations of users, since in the past people have been rather skeptical about such services.

Mobile electronic commerce is another playground for new attractive service offerings. This includes instant payments through the mobile terminal as well as the provision of custom mobile on-line bank account information, mobile brokerage, money transfers, and other financial transaction methods.

Yet another field that has not been exploited much in mobile communications in 2G networks is advertising. This will have its place in the mobile communications environment. By offering general customer information or personalized information (e.g., depending on customer profile or location, or both) similar to television, advertising can reach the consumer or client. Advertising will also cover at least part of the cost of making certain services or information/content available to the consumer. The consumer will (have to) accept that certain information comes with advertisements.

12.6   Standards and new technologies

On the cellular technology and standards side of things, second-generation equipment will still be deployed for many years to come, and many countries are still developing in that respect. Over the same period, the rollout of third-generation systems will take place, starting in developed countries at build-out speeds that are determined by factors such as license mandates, investment and return, planned penetration levels, coverage, and capacity. With the existence of second-generation networks many users will have multimode terminals that work on both second- and third-generation networks—and because 3G standards allow different radio access technologies, this will also be reflected in multimode handset technology. Such a "world communicator terminal" would be appealing, making true anywhere, anytime, anyone mobile communications a reality.

What comes next? First of all, 3G, like 2G, will come in phases. The standards describe basic and mandatory features, but also an abundance of optional ones such as different modes and data speeds at different usage profiles and environments (e.g., stationary versus fast moving).

Since available data rates are also a function of network build-out (coverage, capacity, spectrum availability), mobile operators still have a lot of rollout work ahead. In particular, the ones that operate 2G networks have more options to play with when it comes to making 3G service available. This may lead them to delay 3G deployment, or only deploy in dense areas, while using 2.5G systems as a way to test demand for advanced applications and allowing 3G networks to mature and handset prices to fall.

The ever-existing demand for additional spectrum will not disappear. The need for additional spectrum will have to be met by clearing new frequency bands for common worldwide mobile communications use. It might be expected that fourth-generation systems will emerge and spectrum will need to be found for these. Given the time taken to clear radio spectrum, it would be appropriate to start identifying candidate bands now.

In my opinion, too little is known today to be certain of the technologies and services that will be implemented in the next (fourth) generation wireless communications networks. In general, the new systems and networks will need to use much higher granularity in access network distribution (i.e., smaller cells) for getting higher capacity and coverage and to provide for less transmitted maximum RF output power. This would also better address the issues of RF propagation/radiation and power consumption, in that it will make sure that the minimum necessary RF power for connections is much lower than in today's networks.

We do live in a world of ever more quickly developing technologies in all domains. The fact that in the entire past century it was always true to say that the future has already started gives us exciting prospects for things to come in wireless communications.

CHAPTER

13

Contents

Communications cocktail

Mike Short

"Whatever you do, or dream you can, begin it now. Boldness has genius, power and magic to it."

—*Goethe (1749–1832)*

Over the next 20 years the opportunities to communicate via different means of public and private network access will multiply beyond recognition. To quantify and qualify the key trends over such a period is a challenge, but in order to give guidance to research and development and major capital investments, it is important to be bold.

This chapter starts by examining the current situation through a review of the existing public and private systems. Section 13.4 will look at the key trends, both from the market and the technology perspective. One of the key possible constraints to the future is regulation, and a section is devoted to examining this. The chapter finishes with predictions about the future 10 and 20 years away based on these trends and constraints

283

and suggests that the future consists of a cocktail of different communications solutions integrated to provide personalized applications and instant communications.

13.1 The current communications environment—a mix of public and private systems

While public and private networks are likely to compete and converge at some point in the future, dividing current networks into private and public is a useful framework for understanding the current competitive communications environment.

13.1.1 Public access

The likely public communications access networks we will see over the next 20 years are shown in Figure 13.1. The choice of access will multiply beyond the "spokes" of the wheel since a number of technologies can be added to the wheel (e.g., broadband, ADSL, and VDSL), but particularly since the number of public licenses on each spoke will increase (e.g., PSTN with unbundling of the local loop).

The fixed radio access (FRA), home/personal area network (HAN/PAN), digital TV, and digital audio broadcast (DAB) options may require some technology and cost improvement, but they are likely to play a key

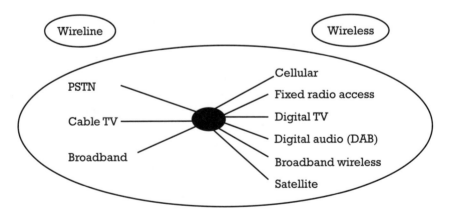

Figure 13.1   Public access.

role in communications in the future. HAN/PANs may also be considered as part of both public/private network systems and convergence since they have a foot in both camps. PANs are likely to be provided using BlueTooth or similar technology.

13.1.2 Private access

The private communications access networks are shown in Figure 13.2 (excluding some of the private walkie-talkie-like systems that do not connect to the PSTN or the Internet).

Historically, most private access has utilized wireline rather than wireless solutions. With new wireless technologies and improved spectrum efficiency, coupled with greater deregulation in the spectrum area, wireless is likely to play a greater role in the private access domain. Cordless systems (home and office PBX) are likely to evolve for voice and data, as are wireless LANs.

Both professional and enhanced special mobile radio (PMR/E-SMR) are the traditional fields of strength in the private wireless fields and serve much of the public safety, utilities, health, taxi, and defense forces with private networks. It was Nextel in the United States that demonstrated how reserved frequencies and technology could be shared between the private and public domains. We will come back to this public/private convergence later.

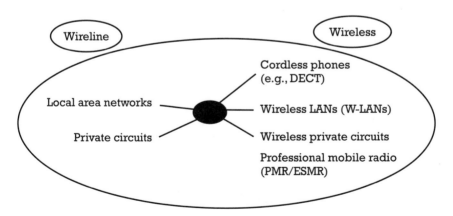

Figure 13.2 Private access.

Private circuits have been available in the wholesale arena for many years, but in the retail arena it has been clear since the 1970s that multinationals and other businesses required voice (and later data) communications, initially between their sites and later with their customers and suppliers. Initially these were provided by both wireless and wireline, and this virtual trend has continued. The impact of this high-availability (always on) service and its typical annual pricing (or flat-fee equivalent) has been felt in most areas of public/private access pricing.

One of the areas most likely to grow rapidly is the use of LAN/W-LAN technology, because of a growing demand for varied data applications. Clearly, speed is important, as is convenience, even if the terminals do not yet command the same economies of scale or volumes we see today in cellular.

New technologies may also arrive; for example, in some quarters ultra-wideband solutions (see Section 5.3) could also claim a place in the next few years.

13.2   Key market trends—personalization, applications, and location

"Any significantly advanced technology is indistinguishable from magic."

—*Arthur C. Clarke (1917–)*

An understanding of market trends is critical, as it is the market that will drive the future of wireless communications. There are many trends that could be addressed here but the most pertinent are listed as follows.

› Increasing personalization;

› Emergence of many new applications;

› Services enabled by location information.

Most of these trends will ignore whether the communications solution is a public or private access solution. They will also be independent of technology, although clearly the underlying bearer has an impact on the customer choice. What will be important, though, will be to apply the three "busy lives" reality checks on any market forecast; they are described as follows.

‣ Is wireline/wireless network access better than traditional means of communicating (hearing, speaking, seeing, smelling, or even writing) without networks?

‣ Does the time per customer per day "on net" make sense? Minutes could also be coupled with megabytes as an additional sanity check.

‣ Is the percent share of gross domestic product (GDP) per head on communications access growing too fast? The percent share of the expenditure on individual products/services rarely doubles in any two-year period.

13.2.1 Personalization

Convergence has been replaced by personalization. Convergence of fixed and mobile networks toward a single communicator has been discussed for many years but so far has failed to occur. Customers' priorities are for more time in their busy lives—more time for access to useful information (not just mail or advertising overload), for their friends and relatives, and for leisure pursuits. This may include working from home, or the ability to contact or be contacted when "away from base." Instead of convergence, these requirements are better met by personalization.

Personalization may be characterized by the following:

‣ *Location:* Voice/e-mails only in the office, home, garden, car, or combinations thereof;

‣ *Time of day:* Perhaps a "do not disturb" voice-mail service;

‣ *Type of network:* This may apply to preferred data rates, cost, or other factors such as QoS;

‣ *Profile or club:* Contact made only if within, for example, the family, golf club, or maternity group;

‣ *Device:* Some may prefer to receive e-mail only on their PCs, but others would be quite happy to review these on their TVs, perhaps shared with others. The scope for wall-mounted flat-screen displays may be limited by the desire to share information.

Personalization as a megatrend will require some flexible thinking by data protection regulators, as the international sharing of databases will be critical to providing a fully personalized service. The personalization benefits may go beyond saving time, to enriching lives and perhaps even carrying fewer communications devices or terminals—after all, who needs a pager, a Walkman, a wallet, and a cell phone, or even a camera and a cell phone?

Access to information (or "my stuff") needs to be adequately secure for business or consumer purposes. This will require better common access methodologies and synchronization techniques. The ability to "push" information to a terminal should also be encouraged, within reasonable personalization constraints.

Personalization is often mistaken for consumerization. Consumer brands such as Apple Computer or Orange PLC may all be part of the communications cocktail, but the brand attributes will need to convey much more in the way of convenience, reliability, and security to add up to a truly personal service. In the next 20 years such attributes should become more international in outlook. There are very few truly international communications brands today, and although constrained by national licensing and perhaps language, this is unlikely to remain the case.

The ultimate vision of personalization is the electronic butler that sifts relevant information based on personal service profiles. We may not all want this vision of Victorian service, but it may help in a low-cost automated way by filtering incoming useful calls and messages from the useless and assisting the completion of outward calls of voice or text.

13.2.2 Applications

Key to shaping the future will be the applications in which people see sufficient value to merit an increase in their expenditure on communications. Wireless applications that are expected to be most widely used include the following:

- Mobile Internet;

- Mobile commerce;

- Mobile entertainment;

- Mobile location.

Before examining these in more detail, other means of providing communications should be examined briefly to understand whether they could also deliver these same applications. The digital TV evolution could add applications from a content-rich direction—this now appears less likely since the television industry is arriving at digital later than is cellular; they are still largely working to national (rather than scaled international) standards, and their interactive plans may also remain dependent on other telecommunications networks.

Similarly, the W-LAN evolution could be quite strong in the mobile Internet/mobile office arena. There are, however, constraints to consider such as spectrum availability, health concerns, and scale economies. Both of these examples, and others, could flourish in growth terms but do not seem to have the international volume growth potential equivalent to cellular. Fixed wireless access is likely to remain slow-growing if common spectrum and standards are not adopted.

Mobile Internet is really only in its first cellular phase, having seen the evolution from SMS to WAP-based Internet displays. In many respects we are entering a "silence is golden" era where textual communications become as important as verbal contact. However, the increasing interest in SMS and the Japanese iMode packet-based content services highlight how much variety there is in textual/visual content on cellular displays. Figure 13.3 covers a GSM standards structure, mapping time of service availability against improved data functionality.

The GSM Association announced that 1 billion SMS messages were sent worldwide in April 1999. This rose steadily to 3 billion in December 1999 and 5 billion in May 2000, and exceeded 15 billion during December 2000. Other equally impressive growth levels can be seen in the Japanese iMode service, which rapidly grew to over 10 million customers in less than one year.

While these illustrations may suggest a consumer revolution in cellular terms, the business needs of mobile have not been confined to voice alone. Access to e-mail is routinely cited as highest on the list of needs when "away from base," as shown in Figure 13.4, which illustrates the current and intended use of wireless data applications and is based on a PMP survey in 1999 of around 1,000 European companies asked about their deployment plans. Figure 13.5 illustrates the main barriers to the growth of mobile data applications in the same PMP survey. It is hoped that with the advent of GPRS or packet radio services in Europe during 2000–2001 that the main barrier of speed will be removed.

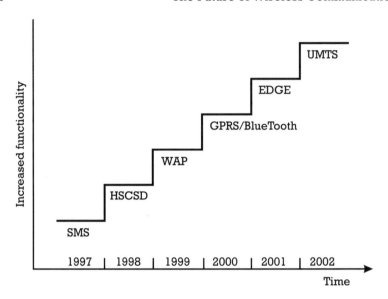

Figure 13.3   GSM standards structure.[1]

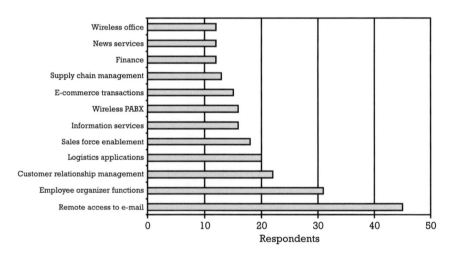

Figure 13.4   Current and intended use of wireless data applications.

---

1. High-speed circuit-switched data (HSCSD) is an enhancement to GSM that enables data rates of up to around 64 Kbps in a circuit-switched, as opposed to a packet-switched, mode.

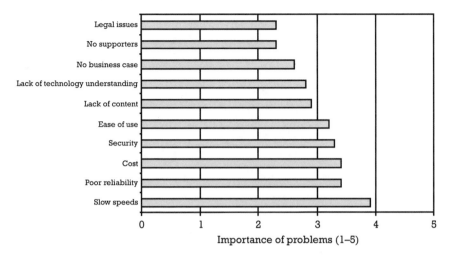

Figure 13.5 Main barriers to growth of mobile data applications.

If additional evidence were required as to the explosive growth of the mobile Internet, the global WAP forecast up to 2005 is provided in Table 13.1, which illustrates the forecast growth in cellular subscribers worldwide, with the relative WAP shipments (nonactive and active) and Ovum's forecasts of related data revenues. It suggests some pessimism of the revenue potential from data while acknowledging that more WAP capability is being shipped into the global cellular market.

Mobile commerce is a natural evolution for wireless devices. Some would say that it will be held back by the reliability of the Internet

Table 13.1

Global WAP Forecasts

			2000	2001	2002	2003	2004	2005
Cellular subscribers	On Jan. 1	'000s	469,085	641,279	818,344	1,002,335	1,187,764	1,369,491
Handsets with microbrowsers	On Jan. 1	'000s	8,888	22,754	117,775	339,133	667,091	993,388
Handsets shipped with microbrowsers	During year	'000s	17,903	107,997	574,723	512,133	758,140	891,789
Active microbrowser users	On Jan. 1	'000s	6,007	15,813	76,469	189,536	322,393	484,317
Voice and data revenues from microbrowser users	During year	$m	8,111	32,556	88,596	159,580	238,732	336,880
Data revenues from microbrowser users	During year	$m	1,024	4,991	14,631	25,819	35,822	46,608

and is certainly constrained by today's terminals. However, we already see examples of mobile banking and mobile ticketing as illustrative of the first stages of this development. It is often said that the difference between e-commerce and m-commerce is like comparing shopping to buying. In a mobile sense, less browsing is expected—it is more of an instant purchase. Some experts even talk of l-commerce or location-based buying from a wireless terminal.

Mobile entertainment will provide new ways to bring leisure into our lifestyles. Downloaded music, images, and videos all require relevant content. Some of it may be private content from personal collections or album selections. From Walkman to Talkman, it is expected that personal communications should be able to combine the best digital files with mobility. Examples we see today are MP3 files and cellular usage offered within the same handset—in time the download of much more content (to appropriate devices) is also expected. Certainly, as displays on wireless devices improve, the current limitations on graphics displays will ease. Even today, fantasy football is being played on mobile phones. Betting and online auctions are also not far away.

Mobile location will offer real advantages to users of wireless communications, and this will be covered more fully in Section 13.2.3.

The variety of applications will multiply in years to come to meet a myriad of customer needs. Wireless has intrinsic advantages of mobility, personalization, and also location. What is clear now, however, is that we are moving from a verbal to a visual world and into a world where text and graphics will play a much larger part in our communications needs and methods.

13.2.3 Services enabled by location information

The increase of worldwide travel, through tourism, business, and other needs, has been made possible through better transport options, but, as with communications, this has led to new demands. Whether contactability or vulnerability, or matching distant needs to remote skills, the need to communicate when "away from base" is increasing. In the next 20 years wireless and cellular, in particular, will play a large part in satisfying these needs. As a framework for this section let's use the "LISA" principle:

▸ Locate;

▸ Inform;

‣ Save;

‣ Alert.

Locating. At present in communications we tend to locate people via a telephone number (for home or office) or via an Internet-based address. With wireless the whole process of searching is based on radio waves that can locate both persons and physical goods and vehicles. The process of finding information often relies on Internet-based search engines. In terms of value, however, contactability and to some extent traceability rely on wireless. Satellite tracking has some advantages on a global scale, but there is little doubt that for person-to-person location cellular location has a head start. This is particularly important if you want to complete a call or a message, but less so if real-time communications is not necessary. However, the principle of locating the person or goods you wish to contact or verify will gain in importance. This relates to busy lives, customer relationships, security, and contactability. There are also downsides to excessive contact relating to potential intrusions on privacy. Coupled with relevant applications and personalization, however, location will remain a very powerful and important service. Already, children are being offered mobile phones by their parents to both assure the parents of their whereabouts, but also perhaps to enhance family communications.

Physical tracking of cars, trucks, and goods in transit is expected to increase as a part of customer service and perhaps for insurance purposes. Wireless provides the best means of doing this. There is a variety of tracking technologies, but most applications developers are looking for economies of scale and geographic reach—hence the initial success of GSM in this arena. The progress of location services will depend on key applications, but if cell sites continue to be reduced in size, it is likely that cellular will remain a key technology for many years to come, offering increased accuracy, perhaps combining with other valuable location-based techniques and databases (e.g., triangulation, angle of arrival, and GPS).

The challenges in the next 20 years are going to revolve partly around addressing all the items we wish to locate. The IPv4 addressing protocol will continue to be enhanced, but it is not suitable for the mobile world without enhanced end-to-end security, more capacity, and a better balanced international outlook. A U.S.-centric Internet will not suit everybody's needs, but the addressing capacity cannot remain constrained by old-fashioned ideologies. Mobility does not naturally coexist with the Cold War era.

Informing relies on relevance and location. It is likely that with increased personalization we will be informed when our favorite sports team or share price has a relevant result—otherwise personalization has failed. It is more than this, however. It will be possible to inform car drivers of the last fuel station before their vehicle becomes empty and inform shoppers in malls of sales of the day (in some cases in exchange for mobile subscription costs). Wireless communications may also enable notified ticketing based on personal profile (airlines, artists) and may even inform you about traffic jams before you get stuck. All these are examples of what could be useful, and many more could readily be envisioned.

The key to which applications will succeed is to determine what people would really like. It could be that they would like to always know of special offers, or items of special interest, or life-enhancing information. Looking forward 20 years, we cannot predict all these requirements, but we can see that the enablers are being put in place, for example, database access, personalization, alert and synchronization mechanisms, security, and push technology. Not all these may be cellular-based—some could be Internet-based—but regardless, location and interactivity will play a critical part in this service.

Saving customers does sound dramatic. It was, however, the E 9-1-1 cellular initiative in the United States that led to technologies for enhanced location accuracy. This is not the right place to review the various technology options (see Section 12.5), but there is little doubt that cellular operators have to support such services—the real question is: how much support?

In the cellular world the licensed operator(s) will work with the emergency and breakdown/rescue services in ways appropriate to each country. Solutions will function in terms of how the emergency or rescue services wish to relate to location and the operators. This may (or may not) be on a commercial basis.

The pricing dynamics, however, also need to suit each market. Not every customer would expect to pay for rescue. Some would expect insurance to cover such services. It would be important to consider not just the technology cost but also the service and partnership costs in each case.

Location-based *alert* mechanisms may seem very similar to the informing mechanisms covered earlier. What is clear, however, is that in machine-to-machine communications some machines are ill served. We already have a culture of servicing a machine when it becomes faulty.

We can foresee a culture of proactive care—preempting faults in people, machines, and cars before they arise. We could be entering a new era of device-to-device (D2D) communications, based on alert mechanisms.

Imagine the vending machine that communicates "empty me" or "fill me" depending on whether cash or physical contents, respectively, are ready for service. Imagine the human health condition that can be communicated for regular checkup—based on an alert by location. Imagine the car notifying service stations that there is something wrong with it.

13.2.4 Summary

Whether it be personalization, applications, or location, wireless will provide many new opportunities in years to come. The current advantages of wireless in mobility and personal communications will be supplemented dramatically. Some of the technology changes in Section 13.3 will illustrate how the mindsets of today need to change in favor of wireless.

Beyond the advantages already mentioned, it is clear that the biggest challenge for wireless demand will be the wireless data business model. The cellular evolution model to third-generation systems, the addition of BlueTooth/W-LANs, and personalization may not be enough to satisfy burgeoning global demand. Some of the remaining challenges will, of course, also be covered by technology and regulatory developments, which we will now touch on briefly.

It is clear that we are moving further into an "instant everything" or "I want it now" world (see Figure 13.6). This will include the "find me, inform me, save me, tell me now" ethos that needs to be adopted if

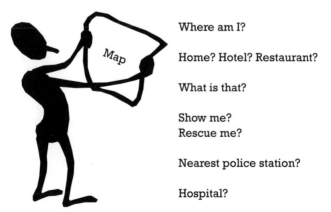

Where am I?

Home? Hotel? Restaurant?

What is that?

Show me?
Rescue me?

Nearest police station?

Hospital?

Figure 13.6 The new world.

technology developments are to be successful. It also should sound a warning bell to regulators who may be standing in the way of such scale innovation and demand.

13.3 Technology trends

The key principles underlying the current technology trends are the following:

> • Globalization and economies of scale in manufacturing;

> • International research and development;

> • Miniaturization.

Most communications networks are still licensed nationally, but their ownership and sources of finance are becoming more international. If the demands of customers become more international (as is expected), network buying power may become more centralized as well. This coupling of demand and supply on a global scale will result in international developments that are more cost-effective when realized on a global scale.

When viewing these trends, it becomes clear that volume will be a key factor behind investor funding and production for the communications industry. Volume will attract skills, innovation, and new customers. It becomes a self-fulfilling prophecy that getting big fast will provide the leading manufacturers with a better chance of survival. Scale without service, innovation, and support will equally cause them to lose their premier division status.

So why miniaturization? Customers will not always want the smallest device, but the ability to deliver the smallest and innovate around it will become very important. It will, of course, have to be coupled with style and convenience. Very few people carry the latest miniature TV because of these factors. If the mobile Internet were complemented with fashion, color, entertainment, and a customer interface that was easy to use, then the market might drive itself. Clearly, cost is a consideration, but confidence does come with economies of scale.

Looking forward 20 years, we see that some of the key technical challenges are the following:

- *Devices:* Memory, display, battery, user interface, connectivity, interoperability, security, compression, operating systems, and application programming interfaces (APIs);

- *Media conversion:* Voice/text/language/formats/translation, plus subsequent repurposing, redistribution, or unification and synchronization of content;

- *Security:* Voiceprint, eyeprint, fingerprint, content firewalls, authentication/encryption, and billing interfaces;

- *Spectrum techniques:* Efficiency, cost, and data throughput.

13.4 Regulatory issues and opportunities

Regulation may play a key role in the development of wireless communications. Most countries view communications as comprising more than one industry. They also often separate industrial, consumer, and spectrum policies. Regulators may face many changes over the next 20 years including the following:

- Public and private access to networks—what's the difference?

- Should content be regulated for broadcast, but not for the Internet?

- How will data protection both protect but also facilitate customer service? Will international rules be compatible? Can databases be combined?

- Spectrum—will it be priced and then traded internationally? If so, when will spectrum databases be opened and published? Will spectrum assignment remain restrictive per sector or become flexible?

- Will different levels of security (in its widest sense) be applied to telecommunications networks but not to other interactive networks?

These questions are relevant as they may fundamentally affect the shape of the communications industry in years to come. We now take some of these in turn.

The private/public systems debate may inhibit combinational services when quite clearly there are already examples of private spectrum being used for public purposes.[2] There are also virtual or semi-private examples in the PSTN world (virtual PBX or Centrex) and in the mobile world (virtual network operators; for example, Virgin Mobile, which uses the One 2 One mobile network in the United Kingdom). This will become far more significant as we see the convergence of fixed and mobile services and perhaps broadcast entertainment and Internet services.

For copyright regulation the approaches are quite different between the broadcast and Internet worlds. To some extent they relate to different business models, but more so to traditional national (versus international) regulations. When, for example, should extracts of content such as music be published for onward sale of the whole?

The implementation of data protection is relatively new and not always compatible internationally. When combining databases from, for example, banking and telecommunications sectors, the rules become less clear. When directories should be encouraged to be open and when they should remain closed is still subject to some confused debate. This confusion will need elaboration before real personalized services grow. It will also need some international consistency early on in order to deliver, for example, calling line identity (CLI)–based services.

Spectrum availability is handled in a variety of ways, and internationally it is not always consistent. The International Telecommunication Union (ITU) may be the final arbiter of treaties and service (sector) assignments, but in practice the national and regional (continental) regulators play a big part. The broadcast and telecommunications regulators vary in their views. The introduction of spectrum pricing and trading is still a relatively new initiative on a national basis. The ability to combine public and private spectrum, or telecommunications and broadcast spectrum, is really quite variable in the way it is adopted internationally. There are no totally consistent approaches to regulating the demand and supply of spectrum, perhaps in contrast to the regulation of land today. The only difference is that usually the databases covering land are a little more open and advanced.

Although satellite TV has some international regulation, broadcast TV tends to be a national licensing matter. Accordingly, the varying nature of how content is regulated and priced is usually a national matter. This

---

2. In the United States Nextel took spectrum and technology originally intended for private use and offered service to enterprises. In the United Kingdom, an operator called Dolphin was rolling out a similar network during 2000.

may include language, advertising, mix of content, and some cultural regulations. This is a far cry from the "many-to-many" world of the Internet, which represents the unregulated extreme. When looking forward 20 years, it is important to consider which of the two extremes is likely to prevail. It is also true that one of the extremes today is more interactive than the other. Internet content may remain largely unregulated, but it is expected that controls for abusive content and "walled gardens"[3] may arise unless it self-regulates. It is also expected that pricing methodologies may apply to speed of access and search, and perhaps also to other service drivers like digital satellite TV.

Security will be particularly important for customer confidence. It is important, however, to separate surveillance from user security. Surveillance may inhibit network use. Trust must also be built up for user confidence in the Internet and, for example, mobile commerce applications. Traceability for inappropriate use of networks also needs some uniformity, perhaps in terms of codes of conduct—this would seem beneficial for user confidence, but it would be wrong to transfer special measures into the virtual world that are not in the physical world. Should the delivery of e-mail be as reliable on the Internet as mail is in the U.S. Postal Service?

An appropriate solution to these regulatory problems would be to unify the regulation of disparate networks into one communications or networks authority, and to pass regulation of all content issues to a separate body. The former would, of course, need to refer to both the international scene and general national competitiveness issues, should the need arise. It remains unclear why broadcast, telecommunications, computing, and Internet networks should remain the subject of separate regulation for much longer, and why equal deregulation cannot be applied.

Overall, the regulatory definitions that apply in various markets should move from a technology or sector basis to one defined around communications.

It is clear that for transportation people base their journey around a bicycle, car, train, boat, or airplane. By analogy, in communications there

---

3. A "walled garden" is a separate private Internet to which users can choose to subscribe. By utilizing a separate network, it is possible to ensure speed and security and to regulate content, although some of the benefits of widespread access to information are lost. Some expect that if the Internet becomes increasingly unreliable, many users will opt for walled garden approaches. Walled gardens are discussed in more detail in Chapter 14.

should be less regulatory distinction between the multiple technologies for network-based communications.

13.5 The future as a cocktail of different communications solutions

In my communications cocktail for 2020, I believe the leading solutions, in order of importance, will be based on:

▶ Personal communications: Cellular;

▶ Home communications: PSTN/ADSL, with growing cordless and W-LAN access;

▶ Office communications: W-LAN;

▶ Away from base: Cellular;

▶ Home entertainment: Digital TV.

This section will elaborate on the above ranking.

Cellular will lead for personal communications for the next 20 years since it has the economies of scale and R&D backing to lead. The benefits of personalization, applications, and location are being added to services all the time and are complemented by the benefits from increased competition.

Overall, cellular will make inroads into the home, but in many developed countries the PSTN will have the added benefit of ADSL and further broadband advances downstream. This, with the benefit of depreciated local loop investment, will generally keep the home in many countries as a communications PSTN domain. The exceptions to this will be in less developed or low-penetration PSTN countries, where wireless (fixed wireless access or cellular) may be the first choice, subject to better regulatory conditions (spectrum, competition, technical) than we see today. In the developed world, however, PC and telephonic access via the PSTN will dominate in the home, with some growth of cordless PC and phone access.

Office communications will undergo many changes in the next 20 years. The most obvious will be out-of-the-office working and then data rate improvements. In terms of the former, contactability with colleagues will become more important, and existing LAN, cellular, and private circuit solutions will be exploited more fully in the next 10 years. However, for on-site use from 2005 the use of W-LANs will complement both the use of PSTN in the home and cellular for personal use. The principal drivers will remain convenience, cost, contactability, and mobility.

When away from base, I expect cellular to dominate based on global coverage and convenience. I also expect that mobility in airports will be met by W-LANs, which become cellular complements as well as stand-alone services. Furthermore, access to the Internet should be very easy using cellular phones by 2002. The ability to download data over cellular and then view this content for easy reading and response will be significantly enabled by BlueTooth by 2002 and be made more practical by the enhancements in display technology expected by then.

I have listed home entertainment separately since I expect bandwidth-intensive communications to be called for over both the (evolved) PSTN and digital TV channels. Home entertainment or "Web pad" viewers may also be used (via BlueTooth) for viewing and listening to Internet or other entertainment content.

13.6 Conclusion

Predictions for 20 years ahead are notoriously difficult. We have to take into account scale, customer perceptions, readiness to change, and more importantly, real need. We also need to understand that dramatic changes in the percentage of GDP spent on telecommunications are unlikely over the short term. It is clear, however, that technology developments and ongoing regulatory liberalization will provide more opportunities. By 2010, we should be in a world of instant communications and coupled with personalization, applications, and location into a world of instant service.

Whatever the outcome, the choice of communications and the availability of information will multiply. We are not heading for a "one size fits all" communications solution but a veritable communications cocktail.

CHAPTER

14

Contents

Mapping the future of convergence and spectrum management

A scenario-based review of spectrum management strategies in the converged era

U.K. Radiocommunications Agency

Late in 1999, the U.K. Radiocommunications Agency (RA), the government body responsible for the administration of radio spectrum, decided that in order to enhance its management of the radio spectrum it would commission a scenario-based study, conducted by NerveWire, Indepen, and Intercai Mondiale, to understand the implications of convergence over the next 10 years, with particular reference to its management of the radio frequency spectrum. The full report on the outcome of the study was published in May 2000 and is available in full at the RA Web site (www.radio.gov.uk). For the purposes of this book, one section from the report is of great interest and has been included here verbatim as one of the key inputs.[1]

14.1 Approach

There is enormous uncertainty about the way in which markets will evolve over the next ten years. Simple forecasting techniques are not appropriate because the past will not be a reliable indicator of the future. For example, ten years ago very few (if any) people forecast the present success of the Internet or mobile phones. Likewise we cannot forecast with confidence the "successes" in 2010.

Scenario analysis deals with these uncertainties by putting forward extreme views of the world, stretching outcomes in a variety of dimensions. NerveWire's Future Mapping® scenario methodology defines a scenario as a sequence of discrete Events that lead to a specific outcome or Endstate. The thought process used emphasises construction of possible major Endstates and then building an Event path to each Endstate. This approach avoids simply extrapolating from current conditions, or working only with scenarios that differ on two dimensions. It also allows for a discussion of what outcomes are desirable, not just which are more likely. Considering the factors leading to a desirable outcome can allow an organisation to create, or at least to influence proactively, the future it prefers.

Events and Endstates are created using data and information collected from interviews, research and our own database of material from previous Future Mapping® projects. In this project we interviewed approximately 18 people within the RA, 10 from other parts of the public sector (DTI, DCMS, Oftel, OFT etc), three from academia, and 25 from industry. We also reviewed other external studies and research.

This gave a rich and customised set of materials for a two-day Scenario Workshop – the pivotal stage of the project. The 27 participants included industry experts, public sector representatives and senior Radiocommunications Agency managers. Four Endstates, a pack of 124 Events and 22 Spectrum Strategy Options were developed by the study team and the RA for the Workshop.

---

1. Crown Copyright is reproduced with the permission of the Controller of Her Majesty's Stationary Office ©Crown Copyright 2000.

Participants were divided into four teams and each team was tasked with defending its allotted Endstate and identifying the key Events and Strategy Options that supported or hindered its Endstate. The teams were encouraged to add new Events and spectrum Strategy Options, as appropriate.

14.2 Convergence Scenarios

In this section, we summarise the four Endstates developed for the world of 2010; distil the key ideas from the workshop teams on each Scenario; review the most important and interesting Events which will influence the speed and direction of change; and present some ideas on how the four scenarios may interact and converge over the course of the decade.

14.2.1 The four 2010 Endstates

The workshop was structured around four Scenarios (Endstates plus Events) that addressed future directions from different perspectives. The four Endstates, labelled A to D, are summarised below.

14.2.1.1 A 2010: The Internet convergence

Theme: *The telecommunications, computing, entertainment, and consumer electronics industries have all converged around open Internet standards and globally available open services. Countries, companies, and consumers that master the Internet economy prosper.*

- Telecommunications, computing, entertainment, consumer electronics have converged around the Internet

- Everything is digitised, TCP/IP has become the norm, there are multiple access platforms, bit transport is a commodity

- The Internet is the basis for interactive TV

- The Internet is part of the fabric of everyday life for consumers, businesses (& machines!); interfaces are user-friendly. Strong brands are highly prized

- Value chains have been shaken up, with many new fortunes, and many failures; choice and customising abound, Electronic Commerce is booming

- Regulation is streamlined and light-touch

14.2.1.2   B 2010: Digital islands

Theme: *Diverse service packages offered through many kinds of networks are successful. Private, closed community networks ("walled gardens") are much more popular than the Internet, which is considered unsafe and a lowest common denominator service.*

- Diversity continues: closed community networks thrive, consumers reject excessive "choice"

- The Internet has fraud, privacy, capacity problems - seen as unattractive, the lowest common denominator

- "Walled gardens" based on leading portals and Interactive DTV are popular for most consumer services; cable does very well

- Geographic clusters, industry-specific nets flourish

- Everything is digital but there are multiple devices and standards in use

- Regulation is little changed, incumbents dig in, open access requirements are modest

14.2.1.3   C 2010: Total mobility

Theme: *Changes in lifestyle and work drive major increases in the use of personal, mobile, converged services. Mobility and untethered operation are an inherent part of the converged world and mobile post-PC devices are the main terminal for Internet and other network services access.*

- Everything is untethered – lifestyle and working habits drive the use of a myriad of mobile, converged services; time, convenience and personalisation are key

- Wireless won Internet control vs. the PC industry; devices have converged around mobile handsets that combine phone, e-wallet, digital ID, Internet access

- WLANs and Bluetooth-style links are common, there are plentiful, affordable public mobile services; wireless fixed access is common, frequency-agile technologies are maturing

- There is a wide range of service providers, some virtual

- Greater industry self-regulation is used on many spectrum and infrastructure issues

14.2.1.4 D 2010: Broadband revolution

Theme: *Tremendous increases in the speed and capacity of wireline networks totally revolutionise communications, information processing, and entertainment. Wireless networks are a poor substitute for broadband optical networks, and are used only when needed.*

- Wireline speeds and capacity have revolutionised broadband communications and entertainment

- Wireless can't match it, and is used only for true mobility and temporary or supporting roles

- Leaps in technology, especially optical fibre, set new standards for quality and reliability at affordable cost

- Uses like Video on demand, Virtual Reality, networked game-playing drive bandwidth demand; the network-centric model prevails

- The electronic-to-optical interface is close to the end user; wireless links are kept short and high bandwidth

- Environmental and health concerns also constrain mobile network and service development.

- The Endstates are not intended to be mutually exclusive; there is a degree of commonality – and indeed one of the main consensus views was that convergence will increase their interaction considerably over the decade.

14.2.2 Workshop conclusions on each Scenario

The ubiquity of the Internet in *Scenario A*, both as an interactive communication route and as a unifying and flexible technology protocol, was felt to be a logical progression. The average user's Internet experience will be transformed by reliable, affordable broadband access from multiple providers and over a range of platforms – most of them more user-friendly than today. The dynamic initiatives by individuals, communities and businesses to exploit the 'Net will drive major growth in traffic volumes –

all digital, of course; there will be some losers, but many more winners. Key enabling Events will include proven Voice over IP technology, the next generation of standards (IP v6), web casting, DTV as an inclusion route and the oxygen of light touch government and a benign regulatory environment.

Scenario B's more diverse and compartmentalised structure for market access, however, seems likely for a large segment of the population – at least through the middle years of the decade. Many consumers seem to value its "trusted brand" blend of convenience, simplicity, and security, versus the "chaos" of the Internet. The attractive business model it offers is already driving many service providers, and the launch of interactive DTV just ahead of widespread consumer broadband Internet access is positive timing. 3G mobile will bring another wave before mid-decade. But, barring major failure of technology or security on the open Internet, a strong trend to A seems likely by the end of the decade.

The richly diverse and truly personalised services enabled by UMTS look to be the key to *Scenario C*. Its fast, always-on Internet access could drive explosive traffic growth, with a variety and real value barely imagined with today's restricted technology. Mobile commerce, bringing a good revenue stream, may prove more of a killer application than entertainment (though young people playing games online may grow really strongly). Tailored Web content should temper the admitted speed disadvantage versus wired links (especially as fibre penetration grows), and Bluetooth-enabled devices and Wireless LANs seem natural winners. This Scenario places the greatest demands on the spectrum managers; new technology will help, but a combination of release (or sharing), reallocation and refarming will also be needed.

The ubiquitous super-high speed connectivity of *Scenario D* seems a natural long-term outcome, linked with the low power pico-cell short hops of *Scenario C*. The Walled Garden phenomenon through the mid-decade will be an important precursor, bringing millions of slow adopters into the fold. Content will be King, and bit transport a commodity; applications will expand, Parkinson-like, to absorb each new quantum of capacity. Service providers are likely to provide the subsidies needed to finance investment and accelerate broadband take-up. Timing was the main area of debate; will 2010 prove a little soon?

14.2.3 Events and their linkage to Endstates

The linkage between individual Endstates and the Events that will facilitate (or hinder) their achievement is an important and valuable element of Future Mapping. In this section we comment on some of the conclusions that can be drawn from these detailed voting patterns.

Consensus Events are good or bad for all/most scenarios; the ones that are also highly likely are good assumptions on which to base a strategy. The following are themes raised by the Consensus Events in the workshop:

> • Low-cost terminals (of many types) are available early, to make it easy to subsidise/give them away to get new subscribers on new services. Government and industry will work together to make them widely available.

> • Ease of Use: Consumers wish for simplification/ease of navigation/trust. Can technology become user-friendly for the mass market? Will voice control, touch screens, handwriting recognition, etc. make it easier to use than keyboards? Better search and agent technology will help.

> • Distance learning is embraced by old and young

> • Multi-player on-line games are very big business

> • Music delivery via the Internet takes off

> • Delivery infrastructure of on-line commerce gets more reliable and customer-friendly

> • Competition in local access spurs development of ubiquitous, affordable broadband services

> • Always-on, always-connected mode of operation for terminals and smart products is commonplace

> • Wireless LANs in the home are widely adopted

> • There continues to be no scientific evidence of health hazards from wireless transmissions or mobile phone usage

Critical Events can either represent *signposts* that indicate a track toward one particular future; or they can be *triggers* or *necessary enablers* in a causal chain. Occasionally an Event may represent a "Fork in the Road" –

with a significant influence on which one of two possible futures becomes more likely. Below, we review the Critical Events under a number of themes.

14.2.3.1 Spectrum availability

- Availability of additional spectrum, re-allocation of spectrum, particularly to mobile services. What about the split between mobile and broadcast? How does the analogue broadcast spectrum get re-used (promotes Endstate C)?

- The Ministry of Defence is able to release or to share spectrum (cost would be less than providing converters for up to 30 million analogue TV sets) (particularly helpful to Endstate C)

- RA frees up/releases more spectrum for:

  - Mobile; in 2 to 3 years operators are likely to seek more spectrum for 3G (promotes Endstate C)

  - Short-range mobile devices on an unlicensed basis (Bluetooth etc.), especially for home networks (promotes most Endstates, because it encourages widespread use of networks by multiple people in the home simultaneously)

  - Broadband wireless tails to the optical broadband network; creates a practical alternative to fibre to the home (promotes Endstate D especially)

14.2.3.2 Government action

- Taxation/regulation/intervention—needs to co-ordinate with major players in the industry, to avoid unintended obstacles to service development (could affect A in particular).

- Government needs to be involved in creating subsidies and incentives to ensure inclusion and universal coverage, working in hand with business. Ensure broadband is extended to less densely populated areas (promotes Endstates A and D especially).

- Government is aggressive in promoting development of socially valuable web content (helps A).

14.2.3.3 Content

- Content Provider strategies: channels, platforms, business models. Will the big media giants go Walled Gardens? Yes, independent publishers get more space, but the big players will shape the environment (these issues were critical to Scenario B)

- People make a *lot* of money on Internet content services (critical to Endstate A)

- Will content originators sell direct via the Internet? Or will they keep their content in exclusive Walled Gardens? (Endstate B versus A issue)

- Will content owners feel that intellectual property rights (IPR) protection measures on the Internet and other on-line services are sufficient? (Endstate B versus A issue)

14.2.3.4 Broadband drivers
These Events were especially critical to the D scenario.

- Fibre into the local loop

- Unbundling of the local loop by 2001-2. Unbundling *does* deliver broadband competition, which is a major driver for the Internet and gets us on the route to a broadband future.

- Development of broadband wireless tails (fixed and mobile) to the optical backbones using pico-cell and other technologies.

- Need for multiple independent Internet sessions into the home; home offices also require higher bandwidth.

- Internet content going to widespread use of full motion images.

- Widespread sourcing of video streams from homes, as consumers connect new digital video cameras to the Internet; broadband streams *from* the home would be hard to handle in some DSL and cable modem architectures.

14.2.3.5 Forks in the road toward the Endstates

- UMTS must be a success to keep confidence in the wireless world scenario (C); GPRS will be an important precursor.

- How will Digital TV (DTV) and the Internet relate? Will DTV be another route into the Internet like wireless is becoming (Scenario A)? Or will DTV be something distinct and separate, i.e. walled off, from the Internet (Scenario B)? Will the simplicity and convenience of interactive DTV win over the mass market? Or will the richness and rapid diversification and innovation of the Internet leave any "island" technology, or business model, in the dust?

- Will the rise of wireless Internet delay the transition to broadband (scenario C versus D)? Will keeping objects small enough to be easily transmitted over wireless links to small hand-helds lessen the need for broadband services? The carriers will lose voice to the mobile operators anyway, so they need to go to broadband Internet access to still have a business.

- If trust is shaken through fraud or security/reliability threats, then walled gardens may become a preferred solution (B versus A). If security is proven and reliable, and trust in the Internet is solid, we move away from walled gardens. Can Internet content blocking work well enough to make it safe for minors?

- If the Internet stocks "bubble" bursts it would slow progress on the Internet and let walled gardens catch up or perhaps feed development of mobile commerce "malls" instead (B or C versus A).

- There will be a lot of great content on the broadband Internet. Twenty big content providers will decide how it all turns out, by how they handle putting their content on the Internet. Will the TV networks make their content available via the broadband Internet? (B versus the rest of the scenarios)

- Will the PC-TV distinction survive the availability of physically unbundled devices (e.g., wall-hung flat screens) with Bluetooth/wireless LAN connectivity (could accelerate decline in B)?

- Degree of acceptance of the open Internet model versus more closed (and controlled) versions of similar technology for business and consumer markets (A versus B). Will businesses subject themselves to fully open markets or will they retreat to traditional value chain structures, or perhaps consolidate vertically?

- Will the Internet become capable of efficient multicast? If so, it will challenge the traditional broadcast networks. Will the architectures of the interactive digital TV networks be sufficiently robust to allow for a compelling user experience relative to what will be available on the full broadband Internet? (B versus A or D).

14.2.3.6 Other critical Event themes

- Globalisation: Will countries become nationalistic again? Would nation-state walled gardens become the norm?

- International strategy on spectrum management is required; lead but make sure they follow closely. No point leading in Europe if everyone else is moving in another direction.

- Mobile data and voice moves to IP, driven by operators and manufacturers, not government (helps A and C).

- Open standards everywhere (helps all but B).

14.2.4 Interaction of the Scenarios

In the final Workshop exercise the participants, in restructured teams, were invited to develop an integrated picture of the future, using the elements of the four Scenarios.

There was a high degree of common thinking in the team presentations, with scenarios seen as variously co-existing and/or progressing from one to another.

The consensus at the end of the workshop could be described broadly as follows:

The Internet scenario (A) is a logical extension of current technology trends and of what early adopters are doing now. The mid-decade years will decide whether walled gardens (B) will supplant the open Internet as the vehicle of long-term choice for the majority of the population; or if they merely provide a key inclusion route on the way to a full Internet World. The wireless world (C) grows very quickly through the middle years of the decade, merges with the Internet, and steals away much of the voice business of the wireline operators, forcing them into the broadband Internet access business. Broadband (D) will come on strong by the second half of the decade; and would be hastened should there be real evidence of health problems from wireless communication.

The figures below (Figure 14.1) capture this evolution of the four Scenarios. The one on the left has a poor showing for walled gardens but a fast ramp from mid-decade for broadband. The one on the right depicts the composite scenario where walled gardens do well for a few years, delaying the development of a new broadband infrastructure, and then fall off. This question of the pace and degree of success for the B Scenario was the key uncertainty.

The Internet will become a diverse inter-related collection of services, aimed at different customer segments that often use different types of terminals. The wireless Internet is related to the older dial-up Internet. The Internet accessed through digital TV sets is different in many ways (more pre-selected/limited) from that accessed through broadband-connected PCs. Owners of content are adept at repurposing it for different audiences and channels.

But the Internet is not the only way that communications is organised and provisioned. Wireless networks offer many services that have nothing to do with the Internet. The broadcast TV networks do not go away in this decade; they will surely still be here in 2010; terrestrial and satellite broadcasting are inherently efficient solutions for one-to-many communication. However, there will be a great deal of overlap and reuse in content between the Internet, interactive DTV, and traditional broadcast TV.

Convergence means that these different worlds of PC-mediated Internet, broadcast TV, interactive DTV, wireless and the next generation of broadband services and terminals will increasingly overlap. The influence of the Internet will dominate. Figure 14.2 depicts convergence in terms of our scenarios:

‣ Today, the worlds of Internet, wireless and the initial TV-centric walled gardens are quite distinct, while broadband is barely here.

‣ Wireless grows very quickly in the next five years and has significant overlap with the Internet, although there will be wireless walled gardens too.

‣ By the end of the decade, the Internet has grown the most and encompasses most of the wireless world and much of what were walled gardens. Almost all of the broadband world will be oriented toward the Internet.

The parallel development of truly mobile wireless services (scenario C) with next generation broadband services (optical cores with high

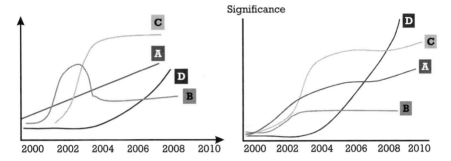

Figure 14.1   Two composite scenarios.

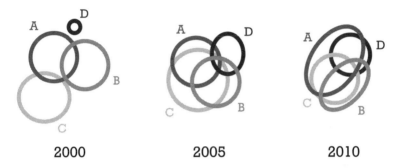

Figure 14.2   A depiction of convergence in terms of the different scenarios.

bandwidth short-distance wireless access – scenario D) was felt by many workshop participants to be the most desirable future. The RA should facilitate the development of truly mobile services of medium bandwidth, as well as more fixed but wireless services of high bandwidth.

May 2000

Prepared for:
Radiocommunications Agency, London, UK
www.radio.gov.uk

Prepared by:

NerveWire, Inc.	Indepen Consulting	Intercai Mondiale
One Liberty Square	Diespeker Wharf	Regatta House
Boston, MA 02109, USA	38 Graham Street	High Street
	London N1 8JX	Marlow SL7 1AB

Contents

Summary of the views of the experts

15.1 Introduction

The summary below was derived by first compiling the key points in each of the different papers and then circulating the summary to all the contributors. What followed was an iterative process where the summary was adjusted until it was agreed upon by all contributors, some of whose views were developed further during this process.

The issues could perhaps be grouped into the following categories:

- Contents, services, and applications;

- The personal communicator;

- The operator;

- Other wireless networks;

- Technology;

- Risks in realizing the future;

- Future scenarios.

We now consider each of these areas.

317

15.2   Content and services

There was a strong feeling that predicting the services that would be provided 20 years from now was extremely difficult. Some felt that it was foolhardy to even try. Others were prepared to make some predictions. In general, the sentiment was that predicting all the services that would be used 20 years from now was definitely not possible. Some of the services could be guessed at, however, and the guess itself would be a useful guide to modeling the future.

Perhaps simplest were the more general remarks. There was widespread agreement that services based around location were likely to be successful. Some observed that the key user needs were to communicate, to organize their lives, and to have fun. Others predicted that users required immediate information. There was also widespread agreement that simply providing mobile access to the Internet was unlikely to succeed—the services would not port well to mobile devices and lower bandwidths. One interesting viewpoint was that the services might first emerge in the home, where high bandwidths might first be available, and then get passed into the WAN. There was also a view that corporate applications such as access to corporate LANs would be the first to succeed because of the higher willingness to pay for such services in the corporate environment. Finally, some expected applications to become more intelligent and to employ agents to search out information from the network. One of the key drivers for this is the increased personalization of the resources, leading to the "electronic butler."

Although a minority of contributors thought that making predictions about specific applications was not worthwhile, others made many predictions, including the following:

- All users will have a VPN profile that will customize the information that they receive and the manner in which they interact with the network. It will take some time for the network to learn the user preferences, but once it does, then it will become an indispensable tool.

- Wireless devices will be highly valued for the security they provide, including interacting with locks and vehicles to identify the owner.

- Key applications include mobile office, Internet, commerce, entertainment, and location.

- M-commerce may be the killer application.

- Distance learning will be embraced by old and young.

- Multiplayer on-line games will become very big business.

- Music delivery via the Internet will take off.

- Video communications might be more successful in a mobile environment (or nomadic environment, depending on data rate requirements) than a fixed one as they could be used to show people where you are (e.g., a picture of the beach).

- VoD will take off in 2005, HDTV in 2015.

- Users will have a complicated set of preferences that sometimes will block out those needing to contact them.

15.3 The personal communicator

The contributors nearly all had a lot to say about the device formerly known as the handset. Here we use the terminology "personal communicator" for a myriad of devices that might include communication, PDA, and other functionality. Almost without exception, all thought that we would become completely dependent on our chosen device and that we would have many variants to choose from.

All saw the communicator as a complex device. It would be a cell phone, a PDA, a personal information manager, and more. It would be multimodal, working on many different types of networks, certainly PANs, perhaps W-LANs, as well as cellular. Finally, it would be possible to tailor a communicator to the user's preference.

The contributors also thought a lot about the manner in which we would interact with the personal communicator. All were clear that dramatic changes in the man-machine interface (MMI) would be required before their future visions could be realized. There were mixed feelings, however, about the likelihood of these changes. Some thought they were somewhat unlikely—at least in the next five to 10 years—and this would severely limit the usefulness of the mobile device for many Internet-type applications. At the other extreme, others envisioned a future with dramatic advances such as 3-D displays and touch sensory input. The "average" opinion was that advances such as speech recognition, foldout displays, and more would emerge in a timely fashion to provide the necessary MMI. As an area of agreement, the contributors felt that there would be many different devices with different MMIs. Users would select the device appropriate to their circumstances. Depending on the types and size of the devices, different MMIs would be possible and the network would adjust the services and presentation of information accordingly. In this manner, the need for dramatic MMI advances would be somewhat softened.

15.4 The operators

All contributors were interested in the future role of the existing cellular operators. Almost all thought that they faced a turbulent time. There was general agreement that today's environment, where an operator was vertically integrated and provided both the network and the customer care, would change. Instead, one type of operator would be the service provider to the customer and others would operate networks. The service provider might then enable the user to roam to multiple different networks—different both in ownership and in type of technology. Some expected these to be based on global brands, or for global brands to emerge to fill this space. Although all agreed that this was inevitable, it was also pointed out that this would not happen overnight and that the current licensing and technology design would hinder its eventual arrival.

There was a general expectation that data would make up an increasing part of the traffic on cellular networks. This will require a number of behavioral shifts, particularly in the area of pricing where current expectation is for virtually free data transmission. This will be solved by providing value-added services that include a fair component for the transmission of bits. Most thought that the fastest way to move to this situation was if the operators restricted themselves to being the "pipes" and allowed innovative application providers to produce high value-added data services.

All this leads to the emergence of new operators. These might be service providers (virtual operators), operators of different types of networks, operators targeted at specific environments (such as the home), or operators of specialized networks such as public safety.

15.5 Other wireless networks

Although most contributors focused on cellular networks, they were also clear that there would be other important networks developed. Most commented on the importance of PANs and W-LANs, and expected that these would become integrated with the cellular network. Some went as far as to comment that PANs might be a real driver of change, enabling new functionality and driving applications.

The home was seen by many as a very important environment. Most expected broadband services to be provided to the home, through a mix of technologies but perhaps with fixed wireless playing an increasing role until such time, if ever, when fiber was delivered to the home. Others thought that the PTT had a major advantage with their depreciated copper link and would be difficult to dislodge. Within the home all the appliances would be linked to a central communications hub. Services provided throughout the home might require total bandwidths of as much as 200 Mbps. Wireless was expected to widely permeate the home, with W-LANs providing connectivity. A few noted that data speeds to the home would likely always exceed data speeds when mobile, and indeed the gap might increase. This might mean that many applications that were viable at home were not appropriate for mobile use.

Fewer contributors were interested in the car, other than to note that they expected the personal communicator to work seamlessly in the car and for the car to provide navigational and other user services. Even fewer were interested in the world of broadcast entertainment. One noted that conventional broadcasting as we know it will slowly change as users increasingly demand personalized content.

15.6 Technological issues

A few technical issues arose. These are not as easy to characterize as the other areas. One contributor thought that there was a possibility of a significant technological breakthrough, discovering some new means to transmit information. Most agreed that communicator complexity would continue to increase as Moore's Law enabled greater processing. A few commented on the transition from circuit switching to packet switching, some noting that IP would need to improve significantly before it would be suitable for the mobile world. One noted that location remained a problem that still did not have a universal solution and might remain a tough problem for the next few years. Another raised the problem of media conversion—modifying content depending on the device and the access mechanism. A few argued that cellular systems would have insufficient spectrum to meet the predicted needs of the users (and hence W-LANs and PANs would be required). However, most did not focus on purely technical matters.

15.7 The risks

Most contributors also wanted to stress the risks that might prevent the emergence of the generally shared vision of the future. Some noted that the risks started immediately. Many thought that 3G was based on unsound assumptions related to demand and applications and that its success was in some doubt. If it was not successful, however, the fallout, particularly in terms of the fall from grace of massive operators, would severely impact the future. Given the earlier comments about the need for change among the operators, this whole area is clearly one of grave concern.

A couple of contributions considered regulation in some detail and came to the conclusion that there were many regulatory risks, including the development of sufficiently flexible standards, the need for "light touch" regulation, the issues surrounding taxation, the problems in protecting a user's data while at the same time making it available for those who had genuine need for it, and the need for governments to intervene to ensure universal coverage and to promote socially valuable content and applications. Those who discussed regulation concluded that if it were poor, that could have a major impact on the industry.

There was concern expressed by some that the Internet itself might become slow and unreliable and that this would result in "walled gardens"—private networks providing better service. This was seen as possibly inhibiting the vision of a single ubiquitous network infrastructure as well as impacting public acceptance. Coupled to this was the concern that security would become increasingly important as users trusted more and more of their "life" to the personal communicator. Any breach of this security was seen as potentially devastating.

Coupled to the general issues surrounding the Internet was the concern, already expressed, that mobile access to the Internet would remain disappointing. It is possible that operator hype would lead to users rejecting mobile data for some time.

Another related issue was that it might be difficult to find the right balance between personalization and global service provisioning.

Finally, a few commented on the social implications. Health hazards and environmental issues remain concerns, although most believe them to be manageable ones. Harder to quantify was the possibility of a social backlash against technology or against the social isolation that it might lead to. Another issue was the cultural divide between communications and content (e.g., TV broadcasting) and the difficulty in bridging the divide. Again, the view was that if dealt with in a sensible manner, these issues could be overcome.

15.8 Possible future scenarios

Perhaps of greatest importance were the future scenarios painted by the different contributors. What was striking was the general level of agreement between them, albeit with different levels of optimism for the future.

There was general agreement that wireless would become the preferred means of communications. It might remain true that more bits will be transported by fixed networks, but more *value* will travel across wireless networks. Perhaps this is somewhat biased by the fact that all of the contributors are connected with the wireless industry, but even allowing for this, it seems certain that voice will be increasingly moved from wired to wireless transport. Wireless will become a very important Internet-access mechanism, perhaps not able to compete on data rates with the home but generating widespread use, especially for immediate and valuable information. Most expected that the amount of information transported across wireless networks would grow massively.

Most contributors touched on the social implications of ubiquitous high-speed communications capabilities with connected mobile devices and broadband pipes to the home. These are likely to result in an "information society" with the information revolution having the potential to be as much of a paradigm shift as the industrial revolution. Those who were prepared to estimate time scales expected this lifestyle to emerge by 2010 and for it to become embedded in the social and technological fabric of society by 2015. Those contributors who chose to illustrate their views with a "day in the life" analysis showed similar expectations of individuals utilizing personal communicators for all aspects of communications and entertainment.

Most contributors gave much thought to the underlying network fabric required to deliver their future scenarios and came to similar conclusions. In general, most noted problems with delivering broadband services over wide area cellular networks, and some conducted analysis to demonstrate that, unless a major technological breakthrough occurred, this service delivery was not plausible. Equally, most noted that W-LANs and PANs could readily provide the high data rates required. From this they deduced that a hybrid type of network was likely, with cellular providing wide area coverage at relatively low data rates and W-LANs providing high data rate islands. One thought that digital TV would remain an important ingredient in the delivery of information to the user. They then concluded that all these delivery mechanisms would need to coexist, with users able to move from one network to another. Some noted that the mechanisms to do this were not yet in place, and would be complex both technically and organizationally, but expected that the huge rewards that would be available would ensure that the problems would be overcome. This conclusion, naturally, reinforced the views that cellular operators were in for potentially turbulent times, although views were mixed as to whether operators would lead this network topology by introducing W-LANs themselves, or whether they would find themselves in competition, or reluctant partnership, with new operators.

Although most came to these conclusions about the networks, there was differing optimism as to the meaning. The majority felt that although turbulent times were ahead, the vision of a better world would enable the problems to be overcome. A minority felt that the turbulence could be too severe to be overcome so readily and might lead to retrenchment and a delayed broadband future.

15.9   Key points raised by each contributor

By way of completing the summary, Table 15.1 provides details of the most significant points made by each of the contributors.

Table 15.1
Contributors' Predictions

Contributor	Prediction
Birchle et al.	Cellular systems will be unable to provide the bit rates needed, and as a result, a mixed architecture with cellular coverage and hot spots will evolve.
Mouly	Data services will not initially have the success currently predicted but will start slowly.
	Wireline or hot-spot systems will be the preferred access method to the Internet.
	Large telecommunications operators will be reduced to the role of a "pipe."
Ojanperä	New services, many of which cannot be predicted today, will emerge being able to use many of the senses.
	Distributed architectures will emerge and dominate.
	A new technology for communications will emerge in the next 20 years.
Oliphant	The user will be in many kinds of different networks at different times, which will be formed around the user as a virtual PAN.
	Wireless access service providers will interface with subscribers, and operators will provide network capacity.
Redl	Enhanced technology will enable better phones with new input and output mechanisms such as video phones.
	Position location will be a key enabling technology.
	3G will roll out in phases; what comes beyond 3G cannot be predicted today.
Short	The future will be a "cocktail" of different communications solutions of which cellular will be the most important
	Personalization, applications, and location services will be added to cellular.
	W-LANs will complement cellular in the home and office from 2005 onward.
RA	The Internet becomes the dominant global backbone for communications.
	Wireless takes much of wired voice minutes, which forces wired operators into the broadband access business.
	Tremendous broadband capacity to the home will be provided by around 2010.

CHAPTER 16

Contents

A discussion of the key issues

"TV is our latest medium. We call it a medium because nothing's done well."

—*Goodman Ace (1899–1982)*

16.1 Summary of the key developments required

In this chapter we deal solely with those issues identified in Part II. We will return to the views of the experts in Chapter 17 where we will integrate them with the conclusions from this chapter.

At the end of Chapter 3, the following issues were identified as potential unmet user requirements. We are now in a position to discuss in more detail in this chapter whether they can be met within the next 20 years, and if so, how they will be met. This, then, will be the key information for the future road map. The key issues were the following.

- Integration with the office and home wireless networks;

- Intelligence in the network such that incoming messages are dealt with appropriately given the time and location of the user;

- How the mobile device will be linked into the computer system for the forwarding of e-mail given the rather different network concepts of voice and e-mail systems;

- The economics of the provision of video communications to wireless devices;

- The extent to which broadband fixed wireless networks will be deployed;

- The extent to which W-LANs will be deployed within private and public buildings and whether they will be widely accessible;

- The file size over the next 20 years and the speed with which users will want to transfer these files;

- Whether technology will be developed to cope with predicted growth in file sizes;

- How to physically communicate with machines—this will most probably be in a wireless fashion for maximum flexibility and may be through BlueTooth technology;

- How humans can communicate with these machines—at present, the devices tend to use speech synthesis to "talk" to the human but accept input through keypad; in the future, more advanced forms of interaction can be expected including speech recognition and artificial intelligence;

- What higher level protocols machines will use to talk to other machines.

16.1.1 Estimation of future traffic loads

In order to perform the analysis that will be required to understand each of these individual items, we will need some estimations of traffic levels so that we can understand the amount of infrastructure required to provide the desired bandwidth. Estimating traffic levels is often difficult since it demands an understanding of the applications that will be used on the network; however, without any estimations, we cannot make progress. Typically, if these estimations prove wrong, they will modify the economics but not the underlying capability of the technology to provide the service. Modifying the economics is likely to change the speed of introduction. So, as a general principle, errors in these estimates will result in the future evolving at a different speed from that which we predict at the end of this section, but not a materially different future. We have already estimated that hard disk sizes will increase around a hundred-fold over the next 20 years and that we might, at least initially, expect communications needs to follow these trends. Here, we look from the bottom up and try to understand how requirements for different kinds of files might change.

Table 16.1 shows a range of applications along with estimates of how the data requirements for each application are expected to change over time. These are based on estimates made by the author and should not be taken as solid predictions of future needs.

Most of the services detailed in Table 16.1 are fairly self-explanatory. Video clips are expected to accompany e-mail and Internet pages in the future. It is also possible that there are many other applications that have not been envisioned, and hence, such forecasts often provide underestimates of demand. However, without understanding future applications and the amount of bandwidth that they will require, no predictions can be made. In considering the reason for change in each of the lines we note that:

- Voice call data rates will fall as voice coding improves. From 1990 to 2000 the basic voice coding rate fell from around 16 Kbps to

Table 16.1
Data Requirements for Each Application

Time Frame	2000		2010		2020	
	File (Mbytes)	Stream (Mbps)	File (Mbytes)	Stream (Mbps)	File (Mbytes)	Stream (Mbps)
Voice call		0.016		0.008		0.004
Video call		0.064		0.064		0.064
Broadcast video		1		1		1
E-mail	0.02		0.05		0.1	
E-mail attachment	0.75		7.5		75	
Internet access	0.1		0.5		1	
Interactive games		0.05		0.4		1

8 Kbps, although error-correction coding and other overheads doubled each of these numbers. We might expect this halving in rate every 10 years to continue.

▸ For video calls, we saw in Section 3.3 that there are two different classes of video for which transmission rates are both likely to remain constant over time. Here we have selected the lower rate, although we note that for transmissions to the home higher rates than this might be preferred by the user.

▸ For broadcast video the same trends as for video calls are likely to hold true. Here we assume that users will be prepared to accept quality at the lower limit of "broadcast quality" transmissions. We also assume, as discussed in Chapter 3, that there will be a growing trend toward personalized "broadcasting," although conventional broadcasting will continue to exist throughout this period.

▸ For basic e-mail, the growth in file size is likely to be limited since basic e-mail typically contains only text. Hence, we show a limited growth in file size here as the text formatting and other capabilities increase. It is in the attachments that we might expect to see the major growth as discussed below.

▸ For e-mail attachments, we might expect the growth to mirror the growth in file sizes, which we expected to be tenfold per decade. In 2000 the average attachment size was around 0.75 MB. Note that in most cases attachments will be delivered to fixed devices, not mobile.

▸ For Internet access we might expect data rates to grow as sites contain more video information. There is little to go on in making predictions here, but we might suppose perhaps a tenfold increase over 20 years.

▸ Interactive gaming is equally difficult to predict. Here we have assumed something close to a hundred-fold increase but perhaps less as compression and video coding technologies will help reduce overall file size.

Starting with Table 16.1, it is necessary to estimate how many times each of the services will be invoked each day. In the case of file transfer, this is the number of files to transfer (e.g., the amount of e-mail sent). In the case of streams it is the length of time that the stream is required (e.g., the number of minutes of phone calls made). Clearly, these numbers will be different for different subscribers. Table 16.2 shows the predicted usage per day for an average subscriber. These are derived as follows:

▸ For voice and video calls the average is somewhere around the 10 minutes per day mark in 2000 (of which all is voice). We expect a large increase by 2010, as the mobile becomes the only voice phone. After this, we expect that increases will be less since users will not want to spend their whole day on the phone. We predict that by 2010, 20% of the total number of calls will be video calls and by 2020 50% of the total number of calls will be video.

▸ For broadcast video we generally expect that people will continue to use cable and TV broadcast for much of their video, only using the mobile or fixed link for individually generated news broadcasts and similar transmissions, in line with the discussion in Chapter 3 regarding the slow speed of cultural change in this area.

▸ We have assumed e-mail traffic will grow tenfold over the next 20 years, again assuming that there will be a natural limit to the amount of e-mail that people will want to receive each day.

▸ For attachments we have assumed around the same percentage of e-mail with attachments as video-to-voice calls, since many of the attachments will be video files, so there will be some correlation.

Table 16.2
Predicted Levels of Usage Each Day for the Different Services

Time Frame	2000		2010		2020	
	Times transferred per day	Minutes per day	Times transferred per day	Minutes per day	Times transferred per day	Minutes per day
Voice call		10		40		50
Video call		0		10		50
Broadcast video		0		10		20
E-mail	10		50		100	
E-mail attachment	1		10		50	
Internet access	10		50		100	
Interactive games		0		10		20

- For Internet access we have assumed a tenfold increase over the next 20 years. Again, people will not want to spend their whole life on the Internet.

- For interactive games we have assumed that although some will use this service heavily, most will not, and so the average will grow only slowly.

Multiplying Tables 16.1 and 16.2 together leads to the data in Table 16.3. Note the transition here from bytes used in Table 16.1 to bits, the more normal measurement for telecommunications traffic.

In order to understand the peak data requirements, we need to convert from daily totals to busy-hour totals. For most operators, graphs of traffic patterns during the day tend to show little use until around 8 A.M. when they rise to a plateau and stay relatively constant until around 5 P.M. They then tail off to much lower levels in the evening. Although this pattern may change as wireless is increasingly used at home in the evenings, assuming that the traffic is concentrated in nine reasonably even hours is adequate for the purposes of the analysis here, and hence the busy-hour traffic can be found as one-ninth of the daily traffic, as shown in Table 16.4.

There are a few points to note about the derivation of data requirements above. First, the somewhat arbitrariness of the approach is rather clear from the accompanying text—in many cases there is very little data on which to base these assumptions, and so, simplistic estimates must

Table 16.3
Average Data Requirements During the Day

Time Frame	2000		2010		2020	
	Mbits/day	Mbits/day	Mbits/day	Mbits/day	Mbits/day	Mbits/day
Voice call		9.6		19.2		12
Video call	0			38.4		192
Broadcast video	0			600		1,200
E-mail	1.6		20		80	
E-mail attachment	6		600		30,000	
Internet access	8		200		800	
Interactive games		0		240		1,200
Total/day		25		1,718		33,484

Table 16.4
Busy Hour and Data Rate Requirements

Time Frame	2000		2010		2020	
	Mbits/hour	Mbps	Mbits/hour	Mbps	Mbits/hour	Mbps
TOTAL/busy hour	3		191		3,270	3,270
Data rate Mbps		0.00078		0.053		1.033

suffice. Hence, we should not base too much prediction upon these numbers; or if we do, we should be aware of their likely inaccuracy.

Next, it is clear that although the coding rates for voice communications might fall, overall the requirements for data are predicted to rise by as much as a thousand-fold over the next 20 years. Although file sizes may only increase a hundred-fold, the number of files transferred might also increase tenfold. A thousand-fold increase seems enormous, but over 20 years it may be a possibility. Note, however, that all this information will not flow to mobile devices (much will be destined for fixed devices), and so, we are not implying that mobile operators will need to increase capacity a thousand-fold.

Finally, it should be noted that these are average data rates, for the average user. Peak data rates, as already discussed, need to be much higher so that file transfer can take place at a rapid speed. Some users will require only a fraction of this data, others much more. Nevertheless, it may be a good estimate with which to move forward.

16.2 Summary of key constraints

The key constraints that were identified in Part III of this book are repeated below. The key technical constraints were as follows:

 ‣ The key technical principles that dominate wireless communications are as follows: Bandwidth is scarce, providing ubiquitous coverage is problematic, the radio channel is hostile, battery power is scarce, and devices generally need to be portable. These drive many of the solutions and constraints in mobile communications.

 ‣ There are a number of laws and observed principles that shape the way that technology evolves and is able to evolve. Moore's Law tells us that processing devices will rapidly become more powerful and less expensive. Shannon's Law tells us that there is a limit to the amount of information that can be transmitted through a mobile radio channel and that there is a maximum gain of perhaps three times, which can be achieved over the capacity estimated for enhanced third-generation systems.

▸ Because of the limited capacity per cell, small cells will need to be deployed in order to provide the overall system capacity required. These bring a range of problems. Small base stations are required; it is estimated that these might become available by 2005 at under $100 each. Backhaul of the information from the base station is needed, which appears to be an ongoing problem throughout the next 20 years, especially for small cells outside of buildings, and this will limit the deployment of small cells somewhat. The complexity of managing a very large number of cells will be steadily reduced within the next five to 10 years with automated network management packages, as will solutions to managing mobility within a small cell environment. As a result of these issues, we expect small cells to become increasingly common over the next five to 10 years, especially within buildings.

▸ A study of the last 20 years suggests that a capacity increase of three times per cell over the next 20 years would be in line with the last 20 and that, compared to computers, the pace of evolution of mobile communication systems can be expected to be relatively slow.

The key constraints imposed by standards, spectrum, and finance were as follows:

▸ If no standard is developed in a particular area, this will provide a severe constraint on the future since standards are critical to enable widespread wireless communications. As the need for a standard becomes increasingly apparent, however, typically one or more bodies becomes prepared to undertake that standardization.

▸ Standards take some four to six years to complete. Hence, this is a severe and important limit on the speed at which change can occur.

- In some cases, standards will be required that span a range of different industries, and the formation of appropriate groups to address this may be time-consuming and lead to the generation of competing standards. This is especially true of machine-to-machine communications and interworking between different networks.

- To gain the bandwidth required, it is necessary to move to increasingly higher frequencies, which results in more problematic propagation. This will result in the spectrum available in large cells being little more than that available today, but in small cells and fixed wireless systems much more spectrum will be available—typically of the order of 100 times more than today. The implication is that high data rate communications in large cells will be expensive but that much higher data rates will be achieved in small cells and fixed systems.

- Higher frequency operation requires more expensive devices. This will keep operation of W-LAN and PAN systems below 10 GHz; but for fixed wireless systems, where a single transceiver is required per household (rather than the multiple transceivers for W-LAN and PANs), this will not be a great penalty.

- Unlicensed operation will become problematic as more devices use the same spectrum. In particular, fixed wireless systems will not be able to use unlicensed bands in the long term, although they may remain viable for W-LAN and PAN operation as long as these systems are designed to intelligently overcome interference.

- Manufacturers and operators will typically have more projects available to them than the resources they have to complete these projects. They will therefore select the most attractive. This will reduce the speed of implementation in some areas. This is likely to slow fixed wireless development until around 2005, and then after that W-LAN and PAN development might be expected to be impacted.

- Environmental issues such as health concerns and land-use issues present serious problems to the operator now, but these are likely to lessen rather than increase as time goes on.

The key issues identified concerning strategy and organizations were the following:

‣ The key manufacturers involved in mobile communications will change somewhat over the next 20 years, but many of today's key players will remain, augmented by a few new entrants, typically coming from other, related, industries.

‣ In the cellular arena, few new operators will emerge but consolidation on a worldwide basis will continue. In the fixed wireless arena there will be a very volatile period until around 2010 when a few global operators will have emerged.

‣ Unless a vision of the future is shared by most manufacturers and operators, the strategic decisions needed to realize that vision will not be made.

‣ It is increasingly difficult for large companies to develop radical new ideas. These will be developed by smaller companies, which will in turn be acquired by larger companies.

‣ Some might postulate that the future is "out of our control" as networks become more intelligent and evolve in unanticipated manners. This would render prediction of the future almost impossible. Fortunately, few would share this belief.

These constraints will be utilized in developing the road maps in Chapter 17.

16.3   Examining the developments required

16.3.1   Introduction

In this section, we examine and discuss each of the developments that we have identified as necessary in order to understand whether they are likely to occur. Based on this understanding, we will be able to predict what the overall network will look like and build a future road map.

16.3.2   Integration of the office and home wireless network with the mobile communications device

By way of reminder, the key issue here is that users would like to be able to use one communications device everywhere that they go, both within the home and office. Today, where there are office wireless networks, they tend not to utilize the same protocols as mobile phones, which results in different phones being needed in the office and the home. These then need different numbers, and integration of service becomes problematic.

It is worth remembering why this difference exists today. It is because different systems are optimized for different tasks. Cellular systems are great for outdoor voice coverage. W-LAN systems are great for indoor data coverage. It is also worth noting that most office buildings and home buildings do not currently have wireless networks. A number of possible paths could be envisioned, which we need to explore and understand:

▸ Buildings could remain without wireless networks, in which case the only communications would be the use a cell phone and reliance on penetration from outside of the building. This might partially meet the needs of the users, although we might expect that capacity would soon become an issue. Buildings would remain without wireless networks if these networks were expensive or had limited utility.

▸ Commercial buildings could gain W-LAN type networks. Hopefully, these would conform to a common standard (probably IEEE 802.11 and its descendants); otherwise, utilization by mobile phones of disparate networks will probably be difficult. Happily, all the signs indicate that standardization will prevail in this area. In turn, these networks could remain data-oriented or could become voice- and data-capable. Again, all the signs indicate that standards such as VoIP will enable voice to be carried over data links so that these networks will develop voice capabilities.

▸ Commercial buildings might gain PAN networks. Again, it is to be hoped and expected that these will be standardized, probably on the BlueTooth standard. As above, it is likely that these will eventually become voice- and data-capable.

▸ Commercial buildings might gain cellular in-building networks. These will probably also double as W-LAN networks carrying data traffic.

▸ In the home, similar options are possible. As discussed in previous sections, it seems more likely that the home will utilize PANs rather than W-LANs because of the predominance of PAN chips in household appliances. In addition to PANs for providing coverage in the rooms, a backhaul technology like HomeRF will be required, but this is not of concern in discussing this particular issue. Instead of PANs, small cellular base stations might be deployed.

The key difference between these various options—excluding the first, which represents the status quo—is that either the internal networks are based on cellular systems or they are based on W-LAN/PAN systems. If they are based on cellular systems, then mobile phones need no modification and seamless communications can be provided inside the building immediately. If they are W-LAN/PAN systems instead, then mobile phones will need to have multimode capabilities to work on these systems. In the case of BlueTooth it is predicted that by around 2003 all cellular phones will have BlueTooth built in. In the case of IEEE 802.11, it is not clear at present that there are any plans for manufacturers to integrate this into cellular phones. We should remember, however, that Moore's Law suggests that adding more complexity to the phones is not really problematic over the time scales that we are discussing here and that adding an extra mode to the phone is highly likely to happen if there are clear advantages for it.

Now we need to ask whether buildings are likely to get any of these networks. We already know that the technology exists today, or is certain to be developed in the near future to provide in-building networks with any of these solutions. The key issue is whether it will be provided at a cost that the users are prepared to pay and whether it will provide sufficient utility. In terms of costs, there have been some recent proprietary studies that have suggested that business users might pay around $40 per month for the capability to use their mobile phone within the office with all the attendant benefits that this brings, especially the potential efficiency benefits that having a single number and being easier to contact provides. At $40 per month, this is $480 per user per year. This is for the provision of voice communications alone and does not include the benefits that the user might perceive from having a W-LAN. We will ignore these additional benefits for the moment, which will mean our deliberations will tend to be somewhat on the conservative side. We next need to understand how many users share a single base station or W-LAN node.

Current W-LAN nodes provide around 1–10 Mbps of data capabilities. We might expect this to extend to 100 Mbps during the next 20 years. Hence, using the data in Table 16.3, we might expect there to be around 100 users or more per node in an office environment. In practice, coverage will probably limit this to perhaps 10–30 users within the coverage area of a particular node. In a home environment there may only be 1–10 Mbps, but this will still support four to 40 users even in 2020. At the lowest level, with 10 users per W-LAN node, the willingness to pay is $4,800 per node per year. This is substantially more than the cost of such nodes, even including the ongoing costs of the provision of backhaul.

From the above analysis it is clear that there is a very strong likelihood that W-LAN and home networking nodes will be economically viable. Hence, we can expect there to be some sort of in-office networks. It is almost irrelevant whether these are provided with cellular picocells or some form of W-LAN since both can provide the user requirements for the same sort of price level. As discussed in Chapter 3, key issues here are likely to include the desire of the building owner to own the network and the desire of cellular operators to enter this market. Around 2000, cellular networks had more momentum than W-LANs to provide in-building networks; however, the added complexity of third-generation systems may result in the pendulum swinging the other way. Because of this uncertainty, we might predict that both will exist. Some buildings will utilize cellular systems, and others will utilize W-LAN systems. Cell phones will be able to interwork with W-LAN systems and W-LAN networks will be able to interwork with cellular networks. What is important is that users will be able to use their mobile phones within the office. One of the main implications of this is that the desktop phone will disappear. It seems incredible for those of us who have had a desktop phone all our working lives, but it will soon be relegated to the same status as the telex machine.

In regards to home networks, it seems clear that PANs such as Blue-Tooth will always be substantially less expensive than cellular networks and will have much greater utility as most electronic devices are designed with BlueTooth chips. The home network, then, if it exists, will almost certainly include BlueTooth, and cellular devices will be able to interwork with this. The key question is whether users will pay for the installation of a BlueTooth network. In Section 4.2 we predicted that for the average home, the total cost of installing a BlueTooth network including backhaul costs might be $600, falling to $300 by 2010 and $180 by 2020. We have no information available on the utility to the consumer of this network, but we would argue that it would be highly useful, offering communications between many consumer devices. Although it is not possible to provide a detailed proof here, it seems almost beyond doubt that homes would not install such networks until prices fell to, say, $500, with penetration rising as the prices dropped further. So, we can assume that by, say, 2005, the home will increasingly have PAN networks with which phones can interact. Individuals will each have their own mobile phones, which will interact with the home PAN network. As with the office, the wired phone will become a museum piece.

16.3.3   Intelligence in the network such that incoming messages are dealt with appropriately given the time and location of the user

This is an important issue for the utility and social acceptability of wireless communication devices. Without appropriate ways to filter incoming messages and information, the wireless device could become so annoying that users will not embrace it to its full extent, resulting in many of the possibilities for wireless communications not being fully developed. Equally, users will be prepared to pay, perhaps on the order of $5 per month, for a service that simplifies their life. Broadly, intelligent filtering is a way to overcome the potential conflict between the clear advantages of having a single device for all communications needs and the disadvantage that most users currently prefer to segment their life into business and personal time and often do not want issues from one area intruding into time earmarked for a different area. To understand the developments required and the progress that might be made in this area, it is worth understanding all the different types of messages and the issues surrounding time and location. In the next 20 years a user might expect to receive the following messages:

- Voice calls;

- Voice-mail messages—of course, whether a caller is asked to leave a voice-mail message rather than talking to the person they are calling will depend to some extent on time and location;

- Video and multimedia calls and the equivalent messages;

- E-mail messages;

- Machine messages from a range of sources such as machines owned by the user or machines with which the user is in proximity;

- Fax messages, although it seems likely that the use of fax will rapidly die as more documents are available in electronic form and issues such as electronic signatures are solved. Hence, we can probably discount fax from this list.

The users will want to segregate their time into different areas such as the following.

▸ Business;

▸ Personal;

▸ Personal urgent only;

▸ Vacation.

They may want to segregate location into these categories:

▸ At work;

▸ At home;

▸ Traveling in the car;

▸ Traveling on foot;

▸ Other specific locations of value to the users.

In such an environment, there will need to be a single service profile stored for each user. This service profile is likely to grow to become a large record. It will include preferences in terms of the way that the calls are being routed. It will need to include some way to distinguish between business and personal calls: perhaps by using a database of known friends, all other calls are treated as business; or alternatively, perhaps the calling party will need to mark their call in some way as personal or business while they are making the call. It may have detailed rules as to how calls are to be treated in particular circumstances, such as how to handle urgent business calls when on vacation.

When a call is made to the subscriber, at some point, somewhere in the process of the call, an intelligent function in the network will need to be aware of the following information.

▸ The service profile of the subscriber;

▸ The current location of the subscriber;

▸ Whether the subscriber has his or her communicator turned on;

▸ Whether the subscriber is currently engaged in another call.

This intelligent function then needs to be able to route the call depending on these factors. The difficulty is that the call may come from a range of different sources in different countries. It may arrive from a simple PSTN line, from another mobile network, from a computer system, from a PAN, or from one of many other sources. PANs are somewhat of a special case as they will interact directly with the communicator, which will need to know how to deal with their messages.

The next problem is the location of this intelligent function. In general, it would seem highly likely that it is located in the network to which the users subscribe for their cellular service because this network will perceive that by providing this function, they will in some way "own" the subscriber. What will need to happen is that the network from which the call originates will need to send a message to the cellular network, detailing the request for a call to the subscriber. The cellular network can then return a message to the calling party detailing the manner in which it should complete the call. This will generally be routing information. Of course, the calling network will need to know how to send information to this intelligent processing function in the first place, but it is likely that, in the same way that routing tables exist in telephone switches and the Internet, there will be routing information available at key points around the world that can guide the routing of the call.

Developing an intelligent network engine to handle the preferences and route the calls is relatively simple. It could easily be implemented on computing platforms available today at a relatively insignificant cost. Users could interact through the Internet to set up their preferences using a full-size screen and simple-to-follow tools. It could be envisaged that such a function would be provided from 2002 onward as an integral part of third-generation networks.

The problems arise in the communications with other networks. In order for such a solution to work, all networks will need to be able to send the intelligent function notification of an incoming call, along with sufficient information for the intelligent function to be able to determine the course of action, and these networks will need to be able to understand the response. Essentially, a common protocol is required between all networks. There are a number of contenders for this. Probably most appropriate are the signaling solutions developed for cellular networks, key among which is probably the MAP function running on top of the signaling system 7 (SS7) protocol stack. If required, MAP plus some of the SS7 functionality could probably run on top of an IP protocol stack. MAP itself would need modifications in order to provide all the information that might be required but it would be a sound starting point. Most cellular switches already utilize MAP or an equivalent IS-41 protocol. Legacy landline switches, however, may not support this signaling. E-mail messages from a computer network will probably not support MAP either but could pass through a gateway where appropriate protocol conversion was performed.

Hence, we might expect that the intelligent function be implemented around 2002, but that initially it be restricted to calls from certain locations—perhaps cellular networks using the same technology. As time progressed and existing switches were updated—perhaps by IP networks of the form discussed in Section 3.2.7—then the interoperability would increase, until 2010 when all networks could handle the intelligent function and this had become an integral part of the service that the user expected from wireless networks. The actual implementation of this function is relatively inexpensive, requiring a computing platform[1] and some appropriate software, and it should be possible to deliver it well within the value that is derived by the end user. Hence, we might expect this service to be a significant source of revenue to the operator.

16.3.4 How the mobile device will be linked into the computer system for the forwarding of e-mails given the rather different network concepts of voice- and e-mail systems

The potential problem here is that e-mail messages are generated on computer networks, which use protocols such as IP and different addressing and delivery mechanism than voice calls, which utilize protocols such as pulse code modulation (PCM) and circuit-switched delivery. As was discussed in Section 3.2.7, some of these differences can be expected to disappear over the next few years as mobile networks move to core IP switching systems and as voice becomes part of a packet transport stream, using, for example, VoIP. Even so, we still need to understand the issues of addressing and the means by which e-mail messages will be formatted to be delivered to a wireless device.

16.3.4.1 Addressing

We have already talked a little about the addressing format used for e-mail in Section 3.2.7. Basically, an e-mail address of the form john.smith@my-ISP.com is translated into a numerical IP address, some of the digits of which are used for routing purposes. Routing information stored around the world enables the message to be routed to a specific location such as the office where the user is typically located. This works well when the user stays in the same place but is less useful when the user is mobile. A few quick calculations show that in the mobile case it is not possible to update all the routing tables located around the world whenever the user moves to a different location since the updating load on the network would be so severe it would swamp all the wanted traffic sent by the users. This problem has been addressed by the mobile IP group who have proposed that the message be routed to the "home" address of the user as normal and forwarded to the actual location of the user. This forwarding process is known as "tunneling" and involves encapsulating the IP packets in larger IP packets, which have a new address on the front of them. When these messages reach the network where the user is actually located, the original message is extracted from the larger message and sent to the user. Apart from the inefficiencies associated with adding an extra layer of header information, this seems to be an appropriate process.

We can now look in more detail at how an e-mail message would arrive at a mobile user. The person sending the message would address the message as normal and it would be sent through the Internet or intranet and arrive at an ISP named by John Smith as his "home" ISP. This ISP would then need to determine what to do with this message. It would consult the intelligent agent by sending it a message in a protocol such as MAP to tell it that a message had arrived for the user and providing some details as to the sender and priority of the message. It is possible that the intelligent agent function actually be implemented by the ISP rather than the cellular operator; however, the problem with this would be that the ISP might have difficulty in obtaining location information whereas the cellular operator always has this information (at least as long as the mobile device is turned on). Assuming that the intelligent agent was in the cellular network then this would return a message with routing information for the e-mail. This routing information might be in the form of another IP address, a "telephone number" to which a data call could be sent, or a request to simply hold the e-mail until the user decided to retrieve it (or until the user's situation changed so that it could then be forwarded to them automatically). The routing information might request just the header, or some other shortened version of the e-mail (this is discussed in more detail below). Based on this routing information, the ISP can then utilize a protocol such as mobile IP to encapsulate the e-mail message and forward it as requested.

This addressing mechanism provides a clear path forward and is technically perfectly feasible. As so often, when discussing these specific issues, the key problem is the development of a common protocol that can be used by both IP networks and cellular (or other wireless) networks. With increasing convergence on IP, it seems likely that the lower layer protocols will be IP-based, but there is a need for higher layer protocols to be able to handle functions such as requesting routing information. The cellular and computing communities will need to work together to develop these protocols and at present they remain rather disparate entities. Only if they share the vision of a common future outlined here, or some similar vision, will there be the impetus to form the appropriate standards groups and develop the necessary protocols. Even the work required to develop the protocols is relatively trivial; it is developing common protocols that is problematic.

Here we postulate that the need for such commonality will become increasingly clear over the next three years and that this will result in standards bodies being formed that will deliver specifications an additional two years later. It will take another two to three years for these to be implemented throughout the Internet and cellular networks in developed countries and for the service to become ubiquitous. Hence, by around 2007 we might expect that e-mail will be fully integrated into all our other forms of communications. Before that date, there will be many proprietary versions offered, for example, by cellular operators who have partnerships with a specific ISP and, as a result, can develop an in-house solution to this problem, but it will not be universally available.

16.3.4.2 Format of content for the device

After the basic delivery mechanism has been provided, there are still issues in the delivery of e-mail content to mobile devices. E-mails increasingly have attachments and Internet links, and in the future we can expect many forms of multimedia files. These will bring about two key problems: first, the size of the file that needs to be transmitted, and second, the manner in which the files will be viewed. Addressing these one at a time:

› *Size of the file.* In the fixed network, the provision of bandwidth is doubling around every nine to 12 months. The result of this is that in the core network bandwidth is plentiful, and for people connected to the core networks by a high-speed service, there are few bandwidth issues. This change in the provision of bandwidth is changing the computing paradigm, making it more appropriate to store services in a central location and for the computing device to retrieve these services whenever they are required rather than store them locally. As already discussed, however, in the mobile arena bandwidth is problematic. It is also variable. A mobile device in the future may have access to only 100 Kbps in a remote location, 300 Kbps in an outdoor location, 1 Mbps in public buildings, and 10 Mbps in their own office. We might expect that some e-mail contain files of 10 MB or potentially much more. Delivery of such a file may be acceptable when the user has a data rate above 1 Mbps but not otherwise. The simple solution to the problem is to determine the data rate that the user has available to them at the current time, based on location information perhaps, and then use preference settings to determine whether to deliver the e-mail or not. A more intelligent method might utilize different compression techniques to change the size of the file depending on the data rate that could be supported. For example, video resolution could be reduced to make the file sufficiently small to be sent rapidly through a low data rate channel. For text files, graphic image resolution could be reduced, or stylized, to reduce file size.

▸ *Manner in which the file is viewed.* There is little point in sending, for example, a detailed Web page to a mobile device with a simple liquid crystal display (LCD). Similarly, some mobile devices will not be able to display video information. Some miniaturized devices may not even have a display and so it might be appropriate to turn text into speech information for delivery to these devices. To do this, the intelligent agent will need some knowledge of the capabilities of the mobile device. This information will probably be exchanged each time the device is turned on. The intelligent agent will then need to be able to reformat content in a manner appropriate for the device. Some of these services are already available; for example, some cellular operators currently offer a service where e-mail is read out to the cellular subscriber who can reply using voice recognition software. In this area, less standardization is required. Each intelligent agent, if it were required, could utilize different methods of compressing the information so long as the resulting file was in a standard format that the wireless device could understand.

From the discussion above it appears that these problems are entirely soluble and that we might expect proprietary solutions by 2002–2003 and unified solutions by 2005–2007. Hence, it seems highly likely that the user will be able to unify the delivery of e-mail and "call" messages. Users will value this unification of messages and will probably be prepared to pay a few dollars per month, or alternatively pay for the airtime required to deliver the messages. This will easily cover the cost of the computing platform and software required to deliver the service, and so, as with the previous issue, we might expect the solution to this problem to turn into a revenue source for the operator.

16.3.5 The economics of the provision of video communications to wireless devices

We have already discussed that video communications is desirable to some people in some situations and it is something that they would be prepared to pay for. Equally, however, video communications will always require greater bandwidth than audio communications (since audio communications is a component of video communications), and hence, the provision of video communications will always be more costly to the network operator. It seems reasonable to assume that the operator will pass this incremental cost to the end user—that is to say that video communications to wireless devices will generally be more expensive than audio communications. In order to understand whether wireless video communications will be a key driver for the next 20 years, we need to understand how much the service will cost to provide and whether users will be willing to pay it. We do not need to ask many questions about the technology since we have already surmised that video transmissions to mobile devices are already possible, so the issue is purely an economic one.

Unfortunately, the questions of both cost and ability to pay are very difficult ones. We discuss some of the pertinent issues and possible boundaries below before drawing some generic conclusions.

16.3.5.1   Cost of provision

The cost of providing a video call depends entirely on the bandwidth required. For the subsequent analysis we will measure the required bandwidth in terms of multiples of the bandwidth required for a voice call. In 3G systems we might assume that around 10 Kbps would be required for voice communications (perhaps voice coding at 4 Kbps, then additional overhead for error protection, addressing, and so forth). For video transmission to a mobile, low resolution is acceptable since it is unlikely that the device would be able to display more than perhaps 200 × 100 pixels—although for video transmission to a laptop, much higher resolution could be supported, but we will concentrate on mobile terminals for the moment. A low refresh rate is acceptable since LCD displays themselves cannot react to fast changes. Using coding standards such as MPEG-4, we might expect a total transmitted bit rate of perhaps 64 Kbps at some time in the not too distant future. Video transmission, then, is likely to require perhaps six times the bandwidth of a voice call. From the point of view of the operator, this means that a single video call would take up bandwidth that could be used for six voice calls, and so, to remain cost neutral, they would expect to charge a video call at six times the cost of a voice call. Some operators may choose to reduce this margin on the basis that not all the capacity would necessarily be filled, and some may adopt illogical pricing models, but in general we would expect video calls to be priced at six times voice calls. This differential may not change too much over the next 20 years—although there will be advances in video coding, there will also be advances in voice coding, which will approximately remain in the same proportion to each other.

The next issue is the actual cost of a voice call over the coming years. Over the last five years, the cost per minute of a mobile voice call has fallen dramatically from around $0.20 per minute to perhaps $0.05 per minute, although today the cost per minute can be difficult to derive because of the bundling of call minutes and other services into a single price package. It seems certain that costs of provision will fall further as the costs of cellular infrastructure fall (although recent trends in auction prices for third-generation licenses may reverse this trend or at least slow the decline). Some have postulated that this decline will result in voice calls eventually being provided for free, with revenue being generated on other services such as ISP provision, advertising, and so on, as is the model in the Internet. If this extreme scenario occurs, then video communications would also be free and would be widely used. However, this seems unlikely. Free provision will result in massive usage—as operators of One 2 One discovered in the United Kingdom—with users inappropriately using their phones.[1] Bandwidth can never be free in the wireless world, and so this model would seem unlikely to work. Perhaps a limited number of free minutes might be provided with subsequent minutes charged for, but in this case the call is not really free. Nevertheless, this illustrates the difficulty in determining where prices will eventually go as operators may decide to subsidize basic call costs from the provision of other services. We could certainly bound the problem by suggesting that costs will be in the range of $0.00–0.05 per minute for voice calls, with a greater probability that they will fall toward the lower end of this range. Beyond that, we would be arbitrarily speculating. This would imply that video calls would fall into the range $0.00–0.30 per minute, with the same distribution of likely values.

An additional complication is that the cost of a video call may vary depending on location. For calls made in indoor environments where bandwidth is relatively plentiful, the cost may be lower than in outdoor environments. We will not dwell on this issue since it just adds additional complexity, but we might note that we would expect a higher percentage of video calls to be in indoor and in urban environments because the tariffs will drive users to this behavior.

16.3.5.2 Willingness to pay

Here we enter into an even more problematic world. The only real way to determine willingness to pay is to offer a service at a range of prices and monitor the take-up. Different individuals will have different willingness to pay, and as a result we would expect, in principle, to be able to plot a chart showing usage of video calls versus price per minute of that call. This is a marketing problem, and marketers utilize a range of tools, including surveys, to estimate what this curve might look like. Here we do not have access to such surveys nor the resources to perform the surveys, and so we must rely on deduction and personal behavior.

---

1. There are anecdotal stories that couples with a phone each would use them as baby alarms, placing one in the bedroom beside the baby and making a call to the other. By using a hands-free capability on the other phone, the required functionality is delivered. In this mode, calls could be many hours long every night (and tie up two radio channels!).

The first point of note is that when cellular calls were $0.30 per minute, there were plenty of people prepared to pay this. Hence, the affordability of video calls is not an issue for at least a subset of cellular users—typically those users who have had cellular phones since the mid-1990s or earlier. Affordability and willingness to pay are different criteria, however. Second, we note a universal human preference to have visual communications wherever possible for some communications, especially to people that are already known (most would not be overly concerned with visual communications to a switchboard operator, for example). This preference means that people will pay more. Next we note that the desire for visual communications will vary. When calling a close relative, that desire may be very strong. When calling a business associate, it may be moderate. When calling a salesperson, it may be very weak. So there will be a different willingness to pay depending on the person called. There will also be a different willingness to pay depending on the place from which the call is being made. If the calling party is walking along a busy street, video communications will generally not be desirable due to the need to navigate.

Clearly, this is a multidimensional problem. Some individuals will be more willing than others to make video calls, and each individual will be more willing to pay more for some types of calls than others. As the price per call falls, we would expect more individuals to make video calls and for those already making video calls to make more of them.

On a personal level, I would be prepared to pay $0.30 per minute for video calls to my relatives today. The company I work for would probably be willing to pay for video business-related calls to clients and to direct reports in the case that it appeared to add value. I am in no way a special case from this point of view, and so it might be expected that much of the population would act in this manner.

In conclusion, we can expect video calls to cost around six times voice calls. Users, however, will be prepared to pay this differential for some class of calls. By the time that third-generation networks are well established, perhaps by 2005, we can expect video communications to become increasingly widespread and the percentage of calls that are video to gradually grow from only 1–2% to perhaps as high as 50% depending on how the tariffs fall over time. By 2010, video calls will be widely used, but we will probably never reach a situation where all calls are video because users will not pay the differential for some classes of calls, or because it will be inappropriate.

16.3.6 The extent to which broadband fixed wireless networks will be deployed

The vision of the future that has been articulated is for broadband communications within the home. An in-home network connects computers, machines, and mobile phones to each other and to the outside network. As discussed in Section 3.3, this may result in users sending and receiving large files to external networks. A key issue will be the speed with which the home is able to send information to external networks. If these connections stay at the existing rate of 56 Kbps or less, video transmissions to the home and the timely transmission of large files will be difficult. We also want to understand the extent to which broadband fixed wireless will be utilized in the next 20 years since this is another important branch of wireless communications.

As discussed in Chapter 3, there are a number of different means by which a high-speed connection to the residence could be provided, including cable, twisted pair, and wireless communications. Here we will consider only fixed wireless. This may overestimate deployments since some of the connections will be made using these other media, but this is not critical for the analysis conducted here.

We know that in 2000 it was possible to provide a high-speed connection to a home using fixed wireless for about $1,500 per connection. This connection would currently be in the 3.5-GHz frequency band and would provide up to around 1-Mbps peak data rates. In the future, data rates in the region of 10–100 Mbps might be required, depending on the file size and the tolerance for download time. One theory, reported in [1], shows that to provide these higher data rates, bandwidths in the region 100 MHz to 1 GHz would be necessary. These bandwidths are typically only available in the frequency bands above 20 GHz, although there are a few examples in lower bands. In higher frequency bands line-of-sight propagation is required, which tends to limit coverage unless multiple base stations are deployed in conventional systems. It is possible that this problem might be overcome in the future using mesh technology, as discussed briefly in Chapter 3. Here we assume that by using some combination of novel technology, by 2005 it will be possible to provide somewhere between 10 and 100 Mbps to as many residential households as require the service.

As before, the question now shifts to one of economics. If it is technically possible, will it be commercially viable? At present, mesh nodes cost somewhere between $2,000 and $4,000 each for operation in these higher frequency bands. These costs, however, like other fixed wireless systems, can be expected to fall over time. There seems little reason why by 2005 the cost of a home node should not be around $1,000. For mesh networks there are few additional costs, although there will be some form of installation cost in the home. We might assume that by 2005 the total cost of supplying the equipment and installing the service would be around $1,300 and that there would be further price erosion as time progressed.

The next issue is to understand what the user would be prepared to pay. At present, users pay around $20 a month for the provision of a basic phone line. Some users pay an extra $30–40 a month for cable modem service providing data rates of the order of 1 Mbps, although like all telecommunications tariffs, these prices can be expected to fall over time. Clearly, users will pay more than the $20 a month for what will be perceived as a better service. Table 16.5 shows the payback period for a mesh system on the initial investment for a range of plausible monthly fees that might be charged.

These numbers do not tell the whole story since the operator will also have cost of operations, marketing costs, call center costs, backhaul costs, and many more costs that might add, perhaps, $10 a month per subscriber to their cost structure. Operators obviously prefer that the payback period is as short as possible, but for some telecommunications projects they are prepared to consider repayment periods of up to seven years. Even were we to assume that there was a monthly payment of $30, of which $10 was accounted for in other costs, the resulting $20 would result in payback periods of between five and six years. At today's price points, payback might be achieved within two years.

As before, the actual price that the market will pay will be different for different users. It might be imagined that if the price point were $20, the same as a typical U.S. rental fee today, all users would opt for this

Table 16.5
Payback Period of Fixed Wireless Systems
with a Range of Monthly Payments

Monthly Payment	Payback Period
$20	5.4 years
$30	3.6 years
$40	2.7 years
$50	2.1 years

service since it would be at no extra charge. However, this would result in a payback period in excess of 10 years once ongoing costs are taken into account. The operators will be seeking a balance that maximizes their revenue and may well settle in the middle of the range quoted at, say, $40 a month for the provision of a 10-Mbps connection with multiline voice service. At this point, we might expect most homes with PCs to take this connection. By 2005 most homes will have PCs—perhaps 60% or higher for developed countries. Hence, we could see a very high penetration of high-speed fixed wireless to the home by 2005. By 2010 we might expect that almost every home in the developed world would have a high-speed connection.

As mentioned earlier, not all these connections will be wireless. Some may be cable, twisted pair copper, fiber optic, or some other mechanism. To a fair degree we are not overly concerned with which mechanism will be used as long as a high-speed connection to the home is provided. There will be some fixed wireless deployments, at the very least in places with no cable and with limited copper. It may prove that fixed wireless is very competitive against these other delivery mechanisms and will take a larger share even where they are present. If we consider that the total number of households in the developed world is approaching 300 million, and assume that 60% might have high-speed connections and that copper, cable, and wireless each take an equal share, then we might see a market for 60 million fixed wireless connections between 2000 and 2005. This market will continue to grow as developing countries gain in wealth and users in these countries also start to look for broadband connections.

Hence, we expect there to be a widespread deployment of broadband fixed wireless by 2005.

16.3.7 The extent to which W-LANs will be deployed within private and public buildings and whether they will be widely accessible

We actually addressed this issue in the answer to the first issue concerning the possible integration of cellular and office networks. There we concluded that W-LANs would be economically viable in office and home environments. If they occur widely here, then they are also likely to be viable in most major public buildings. As we showed in the case study in Section 6.5, economic viability alone is not necessarily a guarantee of success, and in Section 5.4 we noted that W-LANs will have a lower return on investment for manufacturers than some other possible projects. Hence, we might expect that there is some risk that this introduction will not occur, or that it will be delayed. We predict, however, that users will increasingly come to expect this sort of functionality, and in the same way that many hotels now have modem connections and high-speed capabilities, these hotels will lead the way for public buildings to install W-LANs. So we predict that installation will occur widely but that this will be a slow process. Perhaps by 2010 W-LAN installation will start to become widespread, but it might not be until 2015–2020 that it becomes ubiquitous.

The other part of the question was whether these W-LANs could be widely accessed by people entering the building. Since we predicted earlier in this section that there is likely to be more capacity than the "home" users require, it makes economic sense to open the network to "roamers." The only issue is one of compatibility—only if all users have a single agreed standard will this approach work. As discussed in Section 3.2, it looks likely that BlueTooth and perhaps a W-LAN standard, probably based on IEEE 802.11, might become this global standard. So we might conclude that in the course of time W-LANs will become widespread and widely accessible, but that this will take some time and that there will be a fair element of risk associated with it.

16.3.8 File sizes over the next 20 years and the speed with which users will want to transfer these files

This issue was discussed in some detail in Section 3.3, which concluded that it was likely that file sizes would grow by two orders of magnitude over the next 20 years. This was based on the growth in hard disk sizes, which is shown again in Figure 16.1.

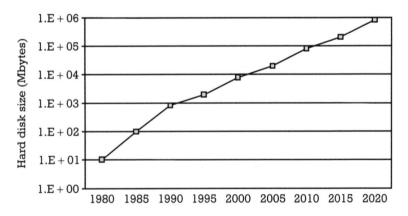

Figure 16.1   Predicted growth in hard disk size (repeated from Chapter 3).

Using the estimate that a user today sends files of up to 1 MB (with a few larger files), we can conclude that this will rise to 10 MB in 2010 and 100 MB in 2020. (There are some who will note that they now transfer 10 MB, but these are generally in the minority.) We can chart the length of time it would take to download files of these sizes at different data rates. Table 16.6 shows the download time in seconds for these two file sizes.

The key issues are to understand the minimum time a user is interested in, the maximum time that they will be prepared to wait, and their willingness to pay more between these boundaries. As always, this is a difficult problem to solve because different people will have differing willingness to pay and because understanding behavior 20 years from now is very problematic. However, as always, we can make some headway using educated assumptions.

First, a user will not be able to "digest" a 100-MB file in a short time period. Even assuming full quality video at today's coding rates, this represents at least two minutes of video. A user is probably prepared to wait 10% of this time, perhaps up to 12 seconds for downloading this file. The same is true for the 10-MB file, for which a user may be prepared to wait around 1.2 seconds. Equally, users will not be willing to wait for longer than the file takes to digest. Hence, 120 seconds for the 100-MB file and 12 seconds for the 10-MB file are probably a maximum that users would be prepared to tolerate unless there really is no other alternative. This actually limits our data rates between quite a narrow bound—something

Table 16.6
Download Time (in Seconds) for Two File
Sizes and a Range of Transfer Rates

Transfer Rate (Mbps)	File Size	
	10 MB	100 MB
1	80	800
2	40	400
5	16	160
10	8	80
20	4	40
50	1.6	16
100	0.8	8
200	0.4	4

in the region of 10–60 Mbps. This is not the whole story because in a house there may be a number of users simultaneously wanting to use high data rate services, so the upper bound may be higher for some buildings. Here, however, we assume for simplicity that the times when their demands overlap will probably be rare.

Assuming that users will expect increasing performance relative to their file sizes over time, we might postulate that users will require around 10 Mbps by 2010 and 60 Mbps by 2020.

The final issue is where users will want to transfer these files to and from. We have already noted that transfer of large files to communicator devices will generally be inappropriate because these devices will not have the resolution to display high-quality video or the storage capacity to handle many of these files. In general, we would expect that these bandwidths would only be required to fixed locations—typically buildings where they would arrive at a home node that could distribute them around the building. Hence, we would predict the need for links to the building to be at this speed and perhaps for distribution within the building also to be at this speed. As Section 3.3 showed, mobile video coding rates are less than 10% of high-quality video, and so we would expect that the need for large files to be sent to mobile devices away from buildings would be less than 10% of these numbers. The exception, as always, is the file transfer to a laptop computer when that computer is out of the building. It is possible to imagine a user watching a full-quality video on their laptop while sitting on a beach away from any building. We should not forget this but we should remember that the demand, and the willingness to pay for high bit rates away from fixed locations, will probably be relatively small.

We have also not talked much about office buildings, or multitenant buildings. Some office buildings already utilize 155-Mbps connections and are expecting this to increase rapidly in the future. Broadly, we might expect that most very high rate connections to office buildings will be made using fiber optic cable and so will not fall within the remit of the wireless communications systems covered within this book. There will be some that do not have fiber coverage, and these may be served by LMDS-type solutions or potentially optical free-space systems. In the future we expect office buildings to need much higher data rates—rates of 100 Mbps to 1 Gbps and perhaps even higher depending on the occupants of these buildings. It seems likely that specialized wired and wireless solutions will be developed to cater to these buildings, probably based on wired and wireless optical solutions. These may be similar to the backhaul technologies used for mesh nodes. These are somewhat specialized systems that are not key to the prediction of future wireless communications systems and so are not discussed in more detail here. Suffice it to say that it seems likely that 1-Gbps wireless systems based on laser optics will emerge by 2005 and will be increasingly widely used by 2020.

16.3.9 Whether technology will be developed to cope with predicted growth in file sizes

We now have a reasonable understanding of the transfer rates that might be required. Specifically, the analysis presented to date suggests that rates on the order of those shown in Table 16.7 will be required.

While the actual rate required will vary from user to user, these are probably reasonable estimates of the rates that might be required over the next 20 years. The key question to be considered in this section is whether technology will be able to provide these rates. We consider each of the different areas separately.

16.3.9.1 Mobile devices

The third-generation systems proposed are able to provide a maximum data rate of around 2 Mbps to an individual user in some specific locations. Here, however, we are mostly interested in outdoor environments. Third-generation systems will be introduced around 2002 and it might be expected at this time that, in outdoor environments, practical data rates will be from 100 Kbps to 1 Mbps with the higher numbers in urban areas. This is substantially greater than the data rates currently offered on second-generation systems, although we have already determined that this additional data rate has not come through additional efficiency, rather through the provision of broader bandwidth carriers. In order to maintain capacity, this requires commensurate increases in spectrum allocation, and this has not been forthcoming. We can note that although the maximum data rate that can be provided may have risen by a factor

Table 16.7
Predicted Transfer Rate Requirements for a Range of Environments

	2000	2010	2020
To mobile devices	100 Kbps	1 Mbps	6 Mbps
To homes	1 Mbps	10 Mbps	60 Mbps
Within homes	1 Mbps	10 Mbps	60 Mbps

of, say, 10 from advanced second-generation systems to third-generation systems, the actual capacity per cell has only increased by a factor of around 2. So, for the same number of cells, on average, third-generation systems can only offer twice the data rate of second-generation systems. We also know that third-generation systems are approaching the Shannon limit for maximum information throughput per cell, so we cannot expect much further improvement in fourth-generation systems to be deployed in around 2012.

This is the key issue. Although it is already possible to produce cellular radio systems with peak data rates of 5 Mbps (for example, the 1XTREME solution proposed by Motorola in March 2000—and this proposal can be extended to third-generation carriers leading to 15-Mbps peak data rates), these rates will require the entire capacity of a cell and will only be available near the center of the cell. Hence, many more cells will be required. If the data rate requirement rises tenfold from 2000 to 2010 but the relative efficiency only increases by a factor of 2 (and even this will not be achieved in some deployments), then five times as many cells will be required. Although there will be some price erosion in the cost of cells, associated costs such the cost of backhaul and site rental will not fall so rapidly, and hence, we might expect that such a network would cost, say, three to four times that of a second-generation network to deploy. This cost has to be borne by the end user, so we can surmise that the end user would have to pay four times as much for the data rate that they would require in 2010 compared to 2000 and probably three times as much again when moving to 2020. Expecting users to pay 12 times as much, in real terms, for what they will perceive as similar utility over the next 20 years when other telecommunications costs are falling will be a difficult proposition to sell.

There is no real solution to this issue. On the one hand we are limited by the laws of physics, and on the other by hard economics. What we can conclude is that the data rates listed in Table 16.7 will, in principle, be available, but they will most likely be so expensive that users will not utilize the rate, and instead seek ways to compress the information or download it in less expensive environments (such as in the building). If we assume that the willingness to pay remains approximately constant, then we might assume that users would actually typically use a maximum data rate of around 100 Kbps in 2000, growing to 200 Kbps in 2010, and 400 Kbps in 2020. These are not "hard" numbers—they will vary depending on the environment and on the current load on the network—but they are probably a reasonable guide to typical data rates for large file transfers, especially streaming files that require that the data rate be sustained for a long period of time.

16.3.9.2 To homes

The situation is somewhat different for rates to homes. This is not because Shannon's Law is any different in the fixed environment; rather it is because much higher frequency bands can be used for fixed transmission and at these higher frequency bands much more radio spectrum can be provided. Operators at 38 GHz have about 1 GHz of spectrum today, whereas cellular operators rarely have more than about 60 MHz. With around 16 times more spectrum, in principle, data rates around 16 times higher can be supported. Hence, without doing much more analysis we might expect to see data rates of 1.6 Mbps now growing to 6.4 Mbps by 2020. Actually, we can do much better than this. The use of directional antennas at the subscriber premise changes the interference perceived on the link, with the result that greater efficiencies can be achieved. Propagation at frequencies above 20 GHz is poor and this further increases the isolation from interference. It is the level of interference that determines the maximum performance, and by making these reductions, further substantial gains are possible, perhaps another factor of 5. Beyond that, we can also note that if operators utilize spectrum in the 40–60 GHz band, another 5–10 GHz of spectrum allocation is possible. Adding all these together, we can come to the conclusion that, in principle, it is possible to provide the data rates required in Table 16.7 in an economic manner.

One key problem, however, is that as we move to the frequency bands at 40 GHz and above, the radio propagation that was advantageous from an interference point of view is disadvantageous from the point of view of getting the wanted signal to the subscriber. As already explained, line-of-sight propagation is essential. The only practical way to achieve this in a suburban setting is either through the deployment of huge numbers of base stations (which is likely to be expensive) or through the use of mesh technology as described in Section 3.2. Hence, we can come to the conclusion that to reach the data rates required to the home in a wireless fashion will require the successful development and commercialization of mesh technology. There are many questions associated with mesh, which have been raised in Section 3.2, so we cannot assume that this commercialization is guaranteed to be successful. Nevertheless, it seems likely that it will be. Mesh technology will probably take until around 2005 to reach a level of maturity where it can be rolled out on a large scale and will probably move slowly up the frequency band, providing approximately the data rates required in Table 16.7 at around the time scales that they are required.

An alternative would be to use wired solutions to the home. We have not dwelt upon these because this is a book dealing with wireless communications, but suffice it to say that, at the moment, ADSL is unlikely to provide data rates above around 8 Mbps to a large percentage of homes. Very high-speed digital subscriber line (VDSL) requires a rollout of fiber optic cable, which will only be viable in some high-density areas, and cable systems will find it difficult to offer these data rates because the bandwidth of the cable itself is less than that at the disposal of the wireless operators.[2] This is not to say that these will not play a role in the next 20 years, but that despite the difficulties of using wireless, there will probably not be a substitute that proves to be simpler and less expensive on a widespread scale while still meeting all of the requirements.

16.3.9.3 Within homes

Section 3.2 shows that the HiperLAN2 and IEEE 802.11a standards had the potential to offer data rates to the end user of around 30 Mbps in today's deployments.[3] In fact, achieving high data rates within the home is much simpler than the previous two cases because the fabric of the building tends to act as a shield, reducing interference from any other nearby W-LAN systems. Furthermore, the short range enables high-level modulation to be used. Given that 30 Mbps is available today, we might expect that 60 Mbps would not be problematic 20 years from now.

The problem is that these two technologies are unlikely to be the technologies of choice for home networking. As already discussed, Blue-Tooth is much more likely to be adopted because of the likely ubiquity of BlueTooth-enabled devices. BlueTooth currently provides a maximum data rate on the order of 1 Mbps. Clearly, it is possible to increase this, although typically at higher costs, and one of the key drivers for BlueTooth has been to maintain a low cost. Indeed, this is the reason for its likely ubiquity. There are three possibilities for the future:

---

2. Most cable systems are designed with a bandwidth of around 750 MHz, whereas a fixed wireless operator at 40 GHz might have over 1 GHz of spectrum.

3. Although the peak data rates are around 54 Mbps, after error correction and other overheads the practical usable rate is in the region of 30 Mbps.

‣ Home networks do not provide the data rates required. This would be problematic as there are already applications that could be envisioned for data rates in the region of 5 Mbps for which users are prepared to pay. Hence, this is probably an unsustainable position.

‣ BlueTooth evolves to provide higher data rates. BlueTooth will almost certainly evolve and these evolutions will almost certainly include higher data rates. Current plans for BlueTooth within the BlueTooth standards body are for a 10-Mbps version, and within the IEEE 802.16 body, a 20-Mbps version.

‣ More than one W-LAN is deployed within the home, perhaps BlueTooth for most devices and also HiperLAN2 for high-speed devices for which the user will pay more to have a higher speed wireless functionality embedded. Another alternative, somewhere between these two, is to utilize the BlueTooth backhaul system, which as we discussed earlier might be based on HomeRF, for high data rate communications in addition to backhaul purposes.

The choice between the last two is really just an issue of economics. The latter option makes device interoperability a bit more difficult, although it is likely that HiperLAN2-enabled devices would also have BlueTooth chips so that they could communicate with devices on the other network. At this point it is too difficult to predict which of these outcomes is most likely, but both will provide the data rates needed, or close to them, and so we do not need to be more specific.

16.3.9.4 Summary
For transmission to the home and within the home we can expect that technology and economics will allow us to approximately keep track with the increasing data requirements of users. In the mobile environment, however, economics will force us to fall behind, changing users' behavior in such a way that they will minimize the transmission of high data rate information when away from an in-building network.

16.3.10 How to physically communicate with machines
In considering this issue, we are interested only in the lower layer aspects of communications, probably layers 3 and below (i.e., we are looking at the specific radio devices and protocols that will be used to communicate with machines). This could be for humans to communicate with

machines or for machines to talk to other machines. In the following two sections we will consider issues related to the higher level protocols and ask how humans will communicate in terms of speech and how machines will communicate in terms of being able to understand what other machines are saying.

First, it seems highly likely that if we do communicate with machines, we will generally do so in a wireless fashion. There will be many exceptions: for example, we will continue to communicate directly with personal computers through a keyboard that is physically attached to the computer. If we communicate with devices like a refrigerator, we will generally want, for example, our personal communicator to talk to the fridge without having to plug it into the fridge or without having to physically connect the fridge and all the other machines into a home network, raising difficult wiring issues and making it difficult to move machines. We have already said that we do not want to predict which machines we may wish to communicate with—we could postulate that the fridge seems sensible (to determine what shopping is required), but there seems little rationale for wireless communications with the toaster. We might decide, however, that some people may want their alarm to communicate with the toaster so that the toaster could automatically make toast by the time that they get to the kitchen. And so on. Instead, it seems better to say that we might want to communicate with almost all machines that are in the house, or in our vicinity when we move outside the house, and that there will probably be many machines, not yet foreseen, with which we will want to communicate in the next 20 years. It is sufficient to leave the issue here as this provides sufficient guidance on range and capacity to allow us to understand the technical requirements of machine communications.

16.3.10.1 Range

We might want to communicate with machines from any range. We might be anywhere in the world when we decide that we want to set up our video recorder to record a key program. Machines in the home, however, will be part of an in-house wireless network. To communicate with the video recorder from afar, we will send a message to the in-home network using perhaps the cellular infrastructure, which can then be relayed internally using the in-home network. The same is broadly true of machines outside the home. Either we will be very close to them or we will be able to use some other network to communicate with the network to which they are connected. For the few machines for which this is not true, they can be equipped with cellular communications devices. We can assume, then, that the range of the communications device within the machine typically will be local, perhaps only within the same room.

16.3.10.2 Capacity

In general, machines do not generate high-bandwidth information. Few people need a video link to their refrigerator. Information from the fridge would be simple text, probably with minimal file sizes. However, there are some important exceptions to this. We might want a video camera to communicate with the in-home network to send live pictures to relatives. We might want one Play Station to communicate wirelessly with another for the delivery of interactive games. There may be other important machines in the future that have similar bandwidth requirements. Hence, to be of maximum value, the physical layer should be able to support at least one medium quality video link of perhaps 1 Mbps per room.

16.3.10.3 Ubiquity

A key issue is that the communications system must be ubiquitous. Each machine must use the same physical layer for communications, otherwise they will not be able to talk to each other. Although communications devices can afford to be multistandard, when it comes to some machines of relatively low value (like toasters), it may be hard to justify the cost of a single wireless system, let alone a multimode system. If the developed world cannot agree on a single standard for this type of communications, it will severely hamper the uptake of wireless communications in this application.

Fortunately, the solutions to all these issues seem to be at hand with the BlueTooth standard described in Section 3.2. It is low-range, has a capacity approaching 1 Mbps (which will probably be increased in future variants), and is the single global contender for this type of application. It is predicted that by around 2003, the cost of adding a BlueTooth chip to any device will fall to about $5. We can expect further reductions beyond that at perhaps 10% per year, so that by 2010 BlueTooth chips will only be around $2.50. With the addition of margins this will add around $4 to the cost of a machine. For machines costing over $200, this will not be likely to be an issue. For machines under this cost, the user must perceive a benefit before spending the extra $4. Four dollars, however, is a relatively small amount of money compared to, for example, monthly expenditure on mobile calls, and so the benefit would not need to be great. Whether there is any benefit will vary from machine to machine:

In the case of the toaster the benefit is clearly very marginal for most users, but on the other hand there is not a significant barrier to placing BlueTooth in almost any machine that would have any sort of benefit from wireless communications.

BlueTooth was starting to be deployed in some applications in 2000 and will rapidly become widespread in communications devices and computers by 2005. It may take somewhat longer before it penetrates devices like refrigerators, perhaps not until 2010 for widespread use in these kinds of machines. At that point users will not immediately replace all their appliances, so it might take until 2015 before we see the majority of homes with most machines using wireless communications. Only toward the end of our 20-year horizon will we really see the ubiquity of wireless communications that could be envisioned. By this time we may well be on the third generation of BlueTooth with enhanced data rates, the ability to extend the range in various situations, increased tolerance to interference, and better networking capabilities. Backward compatibility with the installed base of earlier generation BlueTooth systems will be important. Nevertheless, the overall function of BlueTooth as the physical means of connecting to machines will remain unchanged.

16.3.11 How humans can communicate with these machines

Now that we have established the physical means of communications, we need to consider the higher level issues. Here we consider humans communicating with machines, in the next section we consider machine-to-machine communications.

If humans are using wireless communications, it can be assumed that they cannot physically touch the machine. The options open to them are speech or text. In the case of speech, the human would talk to the wireless device, which would relay the command in some form to the machine. We will return to the form in which it is relayed in a moment. In return, the machine could send information to the wireless device either in the form of speech, to be relayed as audio, or in some form that the wireless device can convert into speech. Alternatively, the human could communicate by pressing buttons or selecting options on the wireless device, much as is done today when remotely activating an answer-phone. The machine could return information in a textual or graphical form, which could be displayed on the device. For example, the machine could send a "picture" of its control panel to the wireless device. The human could then "press" the buttons on the control channel using a stylus in the same way that they can select options today on a PDA such as a Palm Pilot. This has the advantage that the human operates the machine in just the same manner that they do if they are in front of it. Any mix of these methods is possible. The machine might send a picture of the front panel and some speech to say, "Please schedule a maintenance cycle soon." The user might decide to respond with speech or by typing in, "Set maintenance for Tuesday" or by pressing buttons on the image of the front panel. We should expect that in the future users will have a range of options on how they interact with any device, which they can select according to their preferences, their particular circumstances, and the type of wireless device they are using to perform the interaction.

16.3.11.1 Speech recognition

Clearly, communications involving speech requires speech recognition and speech generation. The latter is relatively straightforward and is routinely used today in many computer systems. The former is also available as a software package running on standard PCs. As the user of a speech recognition package will know, currently things are far from perfect. Accuracy is becoming better but there is a long training period during which time many mistakes are made. Speech recognition also tends to utilize most of the memory and processing power of an average computer. Today speech recognition would not be possible on most communication devices because the computation load would be too great. In this area, however, Moore's Law applies and we can expect that in around five to 10 years speech recognition software will be able to run on PDAs, mobile phones, and other complex devices. In another 15 or 20 years it might be possible to embed a speech recognition chip into appliances for only $5–10. The other problem is training. It seems likely that training will almost always be required except for those machines that only have a very limited vocabulary (for example, there are not many spoken commands that one can imagine giving to a toaster!). This could be intensely annoying for the user. A solution to this problem, however, is for all the speech recognition to be performed in the communicator device, which translates speech to text before sending it to the machine that the user is communicating with. Equally, this device could also perform the speech generation, which would provide users with a common feel, and the ability to select a voice that pleased them (as much as any machine-generated voice can be pleasing!). As mentioned above, we might expect communication devices to have the processing power to achieve this function some time between 2005 and 2010. If this proves problematic, they could always call the home PC and send it the speech waveform, getting in return the text, which they could then forward to the required machine. Hence, we can be fairly certain that speech recognition will be possible before 2010 in communicating with machines.

16.3.11.2 Text or pictorial interaction

With respect to the pictorial interaction discussed above, this is possible within the bounds of today's technology. The machine would send a bitmap image to the communicator with some embedded information as to which parts were "clickable." The communicator could then display this on the same display used for video communications and relay "click" instructions to the machine. The machine could inform the communicator how to modify the display whenever a click was received. Protocols for this could be based on the protocols used within Windows to represent objects on the screen and allow parts to be selected.

16.3.11.3 Handling the complexity

Perhaps the final problem to dwell upon briefly is whether humans could handle this. It has been said that fewer than 50% of the population is able to reliably program a video recorder. How will such a population cope with wireless interaction with multiple machines? In fact, it may become easier in the wireless world. The smart communicator could ensure that all machines appeared to work in the same manner; for example, setting the timer on the video was identical to setting the timer on the oven, or on other machines. The smart communicator could guide the user through the operation and remember previous similar operations, performing stages automatically where appropriate. It will take a lot of careful software design, but as devices become more intelligent, interacting with them should become simpler rather than more complex. Some trends in this respect can already be seen. Videos no longer have to have their clock set—they retrieve this information automatically from TV broadcasts. They no longer need to be tuned into different channels—they do this automatically upon setup. They can be configured so that when the play button is pressed the TV automatically changes to video mode. The wireless communicator will enable many more such interactions to be performed automatically. Of course, there will still be some who struggle, but these will be in the minority.

16.3.12 What higher level protocols machines will use to talk to other machines

On the face of it this appears to be a problem of the highest complexity. How can any make of refrigerator talk to any make of PDA and tell it to add milk to the shopping list? We have established that these devices will be able physically to communicate using BlueTooth or similar technology, but what will they tell each other? With humans the interaction is relatively simple because humans have the intelligence to understand the role of a device and the use of the information it is presenting. With machines, this is not the case.

We might imagine the fridge doing the following:

▸ Searching its local radio environment for a device with the tag "PDA—Person who does the shopping";

▸ Sending a message to this PDA of the form "Add to shopping list: milk: 1 gallon: by March 22: preference 2% fat."

So far, so good, although making the PDA aware of whether its owner is the primary shopper in the house is probably a difficult problem in its own right! The key question is who defines the language that these devices will use to converse. Someone has to define the term "Add to shopping list" and program it into all PDAs and all refrigerators (and there are probably other machines that would need to know about this). Next, all possible items that might need to be added should ideally be recognized by the PDA so that it can, for example, help guide the shopper around the supermarket. (Although by this time it may well be that the PDA simply delivers an order to the supermarket over the Internet and the delivery is made directly to the house, but nevertheless the example holds.) The problem is that the number of these possible functions is almost endless and will grow as time goes on and new machines are introduced with as yet unenvisioned functions. The growth itself may be possible to accommodate by issuing new versions of the protocol, perhaps to every home, which can be distributed to all devices connected to the home. However, we return to the key problem of who is going to develop this language in the first place.

This is an issue that the BlueTooth standards committee has understood and has started to address through the standardization of "profiles." By mid-2000, the BlueTooth standard included around 20 profiles covering machines such as PDAs, laptop computers, and fax machines. These profiles started to provide some concept of the sort of communication that the machines might make, although in nothing like the richness of detail that the reasoning above suggests might be required. Fully developing all these profiles will be an ongoing task as more machines add BlueTooth functionality, and we might expect that even the basic set of machines will take two to three years to fully standardize. Developing the software and performing interworking tests once the standard was complete could take another one to two years. Rolling the software out in devices could take an additional year. Hence, we might not expect to see such a standard being widely deployed before 2006. Until that time, machine-to-machine communications will be restricted to machines performing related functions (such as the video and the TV) or machines from the same manufacturer (such as cameras and computers from Sony) using proprietary protocols. If the standard is not developed fast enough, such proprietary protocols could dominate and limit the capability of many machines to communicate with each other in any meaningful manner. Successful development of this protocol will be one of the key drivers toward the future envisioned here.

16.4 Summary of which issues will be solved and by when

Issue: Integration with the office and home wireless networks.
Solution: By 2005 many office buildings will have wireless networks composed either of cellular picocells or W-LANs. They may also have BlueTooth nodes. By 2005 homes will be starting to deploy BlueTooth networks. By 2010 these networks will be widespread. Communicator devices will be multimode and will be able to interwork with all these different types of networks.

Issue: Intelligence in the network such that incoming messages are dealt with appropriately given the time and location of the user.
Solution: Intelligence is provided by an "intelligent function"—a computer database within the network able to perform all these functions. Early versions relating to a single network will be deployed by 2002, but full interoperability with all networks will not be achieved until 2010.

Issue: How the mobile device will be linked into the computer system for the forwarding of e-mail given the rather different network concepts of voice and e-mail systems.

Solution: Standards will be developed to enable this, probably based on mobile IP, although these will not be widely implemented until around 2007. Before this date there will be proprietary offerings from some operators.

Issue: The economics of the provision of video communications to wireless devices.
Solution: Video communications will start to become widespread by 2005. At first they will only be 1–2% of calls but will grow to perhaps as high as 50% by 2010.

Issue: The extent to which broadband fixed wireless networks will be deployed.
Solution: Broadband connections to homes in the developed world will become common to high-end residential customers by 2005 and to all customers by 2010. A high proportion of these connections will probably be wireless.

Issue: The extent to which W-LANs will be deployed within private and public buildings and whether they will be widely accessible.
Solution: They will be widely deployed within most public or private buildings by 2010–2020.

Issue: The file size over the next 20 years and the speed with which users will want to transfer these files.
Solution: We expect file sizes to grow 100 times larger over the next 20 years. We would expect users to have a maximum download time of around 10% of the time it takes to "digest" the file. On that basis, we expect data rate requirements to rise to 10 Mbps by 2010 and 60 Mbps by 2020.

Issue: Whether technology will be developed to cope with predicted growth in file sizes.
Solution: It will not be economically possible to provide the required data rates to mobile devices, which will be limited to 400 Kbps for practical and economic data transfer. Transmissions to the home and within the home, however, will be economically possible at these data rates.

Issue: How to physically communicate with machines.
Solution: This will most probably be in a wireless fashion for maximum flexibility and may be through BlueTooth technology. This will be widespread within high-end machines by 2005 and within most machines by

2010. Because of the replacement cycle for machines, it may take until 2015 before most machines in most homes can communicate wirelessly.

Issue: How humans can communicate with these machines.
Solution: This will be through a mix of speech, using voice recognition, and pictorial displays on the communicator. We can expect speech recognition to be possible on PDAs between 2005 and 2010 and for pictorial methods to be available sooner.

Issue: What higher level protocols machines will use to talk to other machines.
Solution: This needs to be through the development of a common protocol. This is appears to be a complex task requiring the attention of one or more standards bodies. There is some risk that this standardization may never occur, but if others share this vision it is possible that this language will become available around 2006.

Overall: Overall, we can expect some resolution on almost all of these issues over the next 20 years. In some cases, there is some doubt as to the exact mechanism that will be used to resolve the issue, but there are a number of options that would be acceptable. Perhaps the single key area where we do not expect a resolution is in the provision of high data rate communications to mobile devices. This will limit the modes of communications of people when away from buildings, but data rates will be sufficient for video calls and for many other mobile applications.

Based on this analysis and the views of the experts reported in Chapters 8–14, we can now move ahead to develop a consolidated future road map and vision.

Reference

[1]   Webb, W., *Introduction to Wireless Local Loop*, Second Edition, Norwood, MA: Artech House, 2000.

A future road map

"The future belongs to us. In order to do great things one must be enthusiastic."

—*Comte Henri de Saint-Simon (1760–1825)*

17.1 Integrating the views of the experts with those developed here

The experts, whose views are set out in Chapters 8–14, completed their work concurrently with the writing of the text of this book. Given the difficulty in predicting the future, it seems striking that there is similarity, not only between the experts, but also between the work in the other parts of this book. Let us examine the key differences by looking at the different categories developed in the summary section of Chapter 8:

‣ Contents, services, and applications;

‣ The personal communicator;

- The operator;

- Other wireless networks;

- Technology;

- Risks in realizing the future;

- Future scenarios.

Content and services.   The contributors tended to look more at specific applications as opposed to the more general view taken in this book of studying the highest-level view of the requirements of general users. Arguably, the approach taken in the book is top-down, whereas most of the contributors worked from the bottom up. Both are perfectly viable and intelligent ways to analyze the issue. The differences in output are refreshingly small. The contributors brought out some more specific details on the important services, such as location, that had not been considered in detail in the rest of the book. Otherwise, they provided a welcome level of detail on the more abstract thoughts otherwise presented.

The personal communicator.   Here, the agreement was almost perfect. All of us expected complex multifunctional and multimodal devices to emerge. All understood the limitations of the MMI and predicted enhancements to overcome this.

The operators.   The contributors were much more interested in the role of the operator and mostly provided more detail than was presented in Chapter 6. Nevertheless, there is fairly good agreement between their views and those presented in Chapter 6. They are consistent in their predictions and back them with sound analysis. It is likely that their predictions will come to pass.

Other wireless networks.   The experts were less interested in other types of wireless networks than the analysis in this book. Where they did address the issue, they were in agreement with the thesis developed here on the importance of W-LANs and PANs and on the developments likely in the home space. The prediction of data rates of 200 Mbps to the home was even in the same ballpark as the prediction made in here of 60 Mbps—given the fact that this is around a thousand-fold change from today's rates, a factor of 3 in estimation is not overly important.

Technological issues.   There was general agreement here.

The risks.   In line with their increased focus on the operator, the contributors were more concerned with the possible failure of 3G and the fallout that this might produce. A number of other concerns that had been given only limited exposure here were also aired, including security and regulation. Otherwise, there was much similarity.

Possible future scenarios.   Most important was the level of agreement on the future scenarios. Fortunately, and perhaps surprisingly, it is here that the agreement was greatest of all. In this book we have developed a view of wireless becoming the dominant industry, but one where multiple different network types exist in different environments, bridged by multimode communicators. This is precisely the view arrived at by most of the contributors. Even the time scales, where expressed, were in good agreement. There was differing optimism about the future, although the "average" sentiment would align well with that expressed here.

Summary.   All contributors to this book have expressed their concern that predicting 20 years ahead in the fast-moving wireless communications industry is very difficult and prone to a high degree of error—and yet virtually all, despite their differing backgrounds and their working in complete isolation from each other, came to very similar conclusions as to the overall picture. At the start of this work, this seemed a most unlikely outcome.

This would suggest that predicting the future of wireless is not such a difficult task for industry experts as might initially be expected.[1] It further suggests that the prediction expressed here has a good chance of being correct. This is because the future is made by those with vision acting in accordance with their beliefs. Most of the experts who contributed to this book are also leaders in the world of wireless communications. For example, Mike Short was the chairman of the GSM Memorandum of Understanding (MoU) group—the most influential gathering of the world's cellular operators. The contributors have the power to strongly influence the future and they broadly share a vision. They will act in such a manner as to bring this vision to pass. The combination of their actions

---

1. Indeed, the work in Chapter 7 regarding the "pretend" prediction from 1980 and the arguments concerning the impact of standards in reducing the speed of change corroborate this viewpoint.

and the likelihood that other well-informed individuals will reach the same conclusion make the outcome summarized in Chapter 15 highly likely.

In the remainder of this final chapter we bring together all the previous material into one coherent whole. We take the data we have compiled on which services will be developed and combine it with the material from the contributors to deliver a summary picture of the future.

17.2   The overall network architecture

From the earlier chapters in Part IV of this book, we have noted the following key architectural drivers:

- By 2005 many office buildings will have wireless networks, composed either of cellular picocells or W-LANs. They may also have BlueTooth nodes. By 2005 homes will start to deploy BlueTooth networks. By 2010 these networks will be widespread. Communicator devices will be multimode and will be able to interwork with all these different types of networks.

- Intelligence will be provided by an "intelligent function"—a computer database within the network able to perform all these functions. Early versions relating to a single network will be deployed by 2002, but full interoperability with all networks will not be achieved until 2010.

- Broadband connections to homes in the developed world will become common to high-end residential customers by 2005 and to all customers by 2010. Some proportion of these connections, probably a high proportion, will be wireless.

- W-LANs will be widely deployed within most public or private buildings by 2010–2020.

- W-LANs will be deployed in dense urban areas, perhaps by 2005 onwards.

- We expect data rate requirements to rise to 10 Mbps by 2010 and 60 Mbps by 2020.

• Communication will be widespread within high-end machines by 2005 and within most machines by 2010. Because of the replacement cycle for machines, it may take until 2015 before most machines in most homes can communicate wirelessly.

This leads us to the conclusion that there will be a number of different networks in different areas but that these will all be integrated. This is a conclusion that most of the contributors also reached and, hence, one that we can state with some confidence. A diagram of the network might look somewhat like that in Figure 17.1.

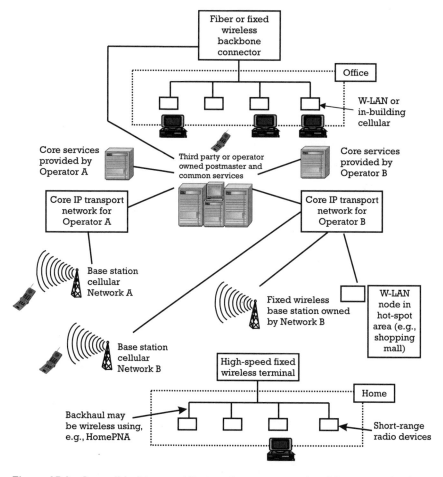

Figure 17.1 A possible future architecture for a converged mobile communications network.

Figure 17.1 may not appear to be a radically different architecture from the one that we have today, but we should remember that the architecture of 2000 was not radically different from that of 1980. Broadly, there are five different networks here:

» The in-home network composed of a number of short-range radio devices, probably based on BlueTooth, communicating with many different devices in the home. These are linked back to a network hub using in-home backhaul, perhaps along the telephone wiring in the home or perhaps wirelessly, using a technology such as HomeRF. The network hub is linked to a high-speed connection into the home. When in the home environment, the mobile phone utilizes the BlueTooth standard, and the information passes back though the high-speed network, like the fixed wireless network, to the core IP network and onto the cellular network;

» The in-office network, which might be in addition to the office Ethernet or might form the only network in the office. This might be based on W-LAN standards such as a development of the IEEE 802.11 standard. The office network is linked via a high-speed connection from the office into the core IP network. This connection is more likely to be a wired connection, probably fiber optic, to provide very high data rates, rather than the connection from the home, which is more likely to be wireless;

» W-LAN–type systems deployed in dense areas such as public buildings, malls, and airports, to provide public service;

» The cellular network, with a similar architecture to today's networks with the exception that the backbone is IP-based using distributed intelligence, as discussed in Section 3.2;

» The fixed wireless network, looking very similar in structure to the cellular network and possibly sharing parts of the infrastructure and core network.

Interworking between these networks is achieved in two ways. Mobile terminals are multistandard and are able to communicate with the cellular, in-office, public W-LAN, and in-home networks. The in-office, public W-LAN and in-home networks are able to recognize the presence of a mobile phone and route information back to the cellular network such that services can be provided in a seamless manner. There is no need for interoperability with fixed wireless networks as these only communicate between fixed points.

The second level of interoperability is the common core network, which uses an IP protocol and a common postmaster function. This ensures that when any device registers on any network, information can be sent back to a common point, perhaps owned by a third party or a cellular operator, and that common services can be invoked.

It is also worth saying a little about broadcasting, a topic we have dealt with only in overview. The work in Chapter 3 suggested that broadcasting will change somewhat over the next 20 years. We expect most people to use a combination of personal "broadcast" services, perhaps related to news and hobbies, and conventional broadcast for events such as sporting events or for simple relaxation. The personal broadcast services will be delivered across the architecture shown in Figure 17.1, from a content server connected to the Internet through whatever channels are appropriate depending on the user's location and preferences. Conventional broadcast will be delivered in a manner similar to today's delivery, either terrestrial broadcast, cable, or satellite. We might expect, however, that terrestrial broadcast will decline as users opt for the higher number of channels available on other media and as the value of radio spectrum encourages terrestrial broadcasters to seek alternative broadcast methods. This trend in itself might be significant: Complete cessation of terrestrial broadcasting would liberate some 400 MHz of spectrum in the 450–850 MHz band in many countries, which could be utilized for cellular applications.[2] This trend, however, will be slow. It will take a long time—at least a generation—for users to change their behavior from watching TV in a broadcast mode on the family television to moving to a personalized mode, and those slowest to move are also likely to be those slowest to change from terrestrial broadcasting to satellite or cable. Forcing the shutdown of terrestrial broadcasting is unlikely to win many votes for politicians! Furthermore, in some countries such as the United Kingdom, digital terrestrial broadcasting is now being rolled out in these bands—consumers buying televisions for digital reception over the next few years will do so in the expectation of 10–15 years of life.

---

2. In the United States, some spectrum in the 700-MHz band currently used by TV broadcasters is planned to be auctioned in 2001. There is unanimous expectation that the winners of the auction will be cellular operators looking for additional spectrum to deploy 3G technology.

It is our view that even by 2020 the existing means of broadcast will still mostly be in place and utilized. Individuals will supplement their broadcast viewing with personalized "infotainment,"[3] but these changes will not have significant impact on the types of wireless communications systems and services that we have forecast here. Therefore, our less detailed treatment of broadcasting is appropriate for this book, and the omission of the explicit images of wireless broadcast systems from Figure 17.1 is an appropriate simplification.

17.2.1 Is there a role for the operator?

One current school of thought suggests that some of the developments charted above will result in the importance of the operator being diminished, if not disappearing altogether. This may seem unlikely and, in any case, not immediately obvious from the previous discussion. Much of the basis for this concept stems from the potential advent of mesh networks.

As already discussed, mesh networks may be one of the key approaches in bringing broadband access to the home. Mesh networks have the property that they are somewhat self-configuring, and because the requirement for base stations is reduced to a need for some mesh nodes to be connected to a high-speed wired connection, then the conventional role of the operator is much reduced. If the mesh network is implemented in unlicensed spectrum, it is possible to imagine a situation where (a little like CB radio) individuals buy mesh nodes and put them on their homes and join a "community mesh" that has no owner or operator. Hence, in this first stage the need for the operator is removed from the broadband fixed network. Advocates of this view claim that this would result in lower cost communications (or, in most cases, free communications once the initial investment in the mesh node has been made). However, there are a number of difficulties with this view:

> ▸ All mesh networks suffer from a "start-up" difficulty. When there are only a few subscribers, there is insufficient subscriber density to form the mesh. At this point, there is no incentive for other subscribers to join the mesh as they will not be able to communicate.

---

3. "Infotainment" is a term formed from the amalgamation of "information" and "entertainment" on the basis that these categories will become increasingly mixed in the future.

Operators can solve this problem by the strategic placement of seed nodes. In an operator-less environment, there seems to be no immediate solution to this problem.

▸ Although it is possible to use unlicensed spectrum, as discussed earlier, this can result in unexpected interference. If all homes started using the same unlicensed band in a mesh configuration, it is likely that capacity would fall as interference rose. Of course, the spectrum regulators could note the trend to operator-less operation and grant further unlicensed spectrum, but this will be time consuming and capacity will suffer in the interim.

▸ Mesh networks require wired access points to connect into other networks. These will require gateways and other expensive equipment and high-speed leased line connections. There will be no incentive for an individual to provide these access points; hence, an operator will be required to step into this arena. It would be possible for the operator to just provide the access point and not the mesh network, but they would then need to charge all subscribers for the use of the access point and so, in effect, have already become a partial network operator.

These difficulties appear to be major ones. Given that there seem to be few ways to solve them, and the role of the operator is very much one that is accepted and implicit to today's telecommunications, it seems unlikely that there will be the innovation and competitive forces necessary to bring about the demise of the operator in the fixed access arena.

In other areas the demise of the operator may be a little more likely. Within the home there is clearly no need for an operator since we expect the user to invest in BlueTooth nodes or other similar forms of communications. The same is true in office buildings where we expect W-LANs to be deployed. It is also true for ad hoc networks formed, for example, in airports as users get close to certain pieces of equipment and form personal area networks. Some postulate that there may be enormous numbers of W-LANs deployed in public areas to the extent that cellular networks become more of a backup. There may be some merit in this—it is likely to vary across regions and depend on the user density.

For wide-area cellular communications, however, the role of the operator still appears to remain a key one. Although mesh networks formed by mobiles could be anticipated, there are immense difficulties associated with traffic management, formation of meshes in remote areas, quality of service as the mesh changes on a second-by-second basis, and much more. It is very difficult to see how cellular communications of the sort that we have come to utilize today could be delivered without an operator.

In summary, the complete demise of the operator does not seem to be something that can be logically predicted based on the evidence available to date. Certainly, however, the role of the operator will change with some operators becoming just network managers and others just service providers. Some operators will find this transition difficult and may go out of business, or more likely they will be acquired by others. In some areas the role of the operator may be reduced, in others it may increase, but we expect that there will still be operators in 2020.

17.3 A future road map

Taking the work in the previous chapters and the views of the experts, we can start to pull together a road map.

17.3.1 2000–2005

17.3.1.1 Key events

- Office buildings will increasingly have W-LANs composed of technology based on either BlueTooth, cellular piconodes, or W-LAN standards.

- W-LAN coverage will start to appear in airports and some other public hot spots.

- Early versions of the intelligent postmaster function will be deployed in some networks, but these will typically be restricted to devices operating on one network and will not be able to work across multiple networks.

- Early third-generation cellular networks will start to roll out.

- Cells and networks become self-planning and optimizing, reducing the management burden on the operator but increasing the software complexity delivered by the manufacturer.

17.3.1.2   Network structure and user capabilities

At this point the structure of the network will be as shown in Figure 17.2. Looking at this diagram, we see first that there are no mobile phones communicating on anything other than mobile networks—users still have different phones for their office, home, and mobile usage. Some offices may have integrated mobile phone systems, but these will be relatively rare. Hence, users are still some way from the unified functionality that they desire. Each cellular and fixed wireless operator offers its own postmaster, but these cannot interwork to any real extent with other networks, offering only a part of the functionality required.

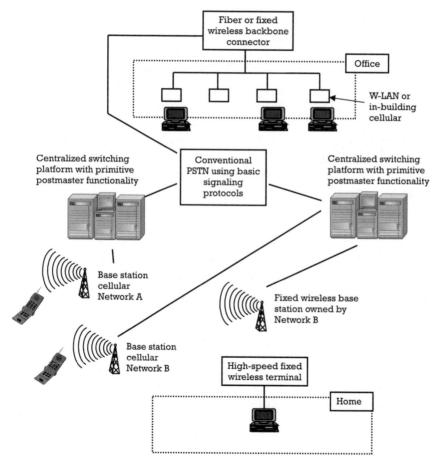

Figure 17.2   Predicted network architecture in 2005.

Next, the core network functionality is hardly any different from that utilized in 2000. Most core networks are still based on conventional centralized switching platforms using signaling protocols such as SS7. Some limited Internet integration and third-party service provision will be provided, and new third-generation operators may be utilizing IP-based core networks, but the huge installed legacy base of Class 5 switches and MSCs will still carry the majority of the traffic. This will make it very difficult to connect different access pipes (such as cellular and fixed wireless) to the same core network and will severely limit the capability to have a single number and phone. The point of interconnect between many of these different networks will remain the PSTN or other privately owned networks. So, for example, linking into the office network while roaming with a cellular phone will still require a "call" to a landline number linked into a modem bank, passing through the PSTN.

In the home, little has changed from 2000, with the exception that many more homes will have high-speed connections through fixed wireless, cable, or ADSL, into the Internet. Some homes will have implemented in-home networks so that multiple computers can be connected to the Internet, but these will generally be wired, or if wireless they will have limited capability to communicate with any other devices.

Not shown here is the gradual penetration of BlueTooth during this period, which results in the formation of some ad hoc networks particularly between mobile phones and computing devices. This will allow automatic retrieval of e-mail using GPRS or third-generation equivalent services.

For the manufacturer, the key changes by this point are the increased usage of W-LANs in offices, the need to integrate BlueTooth into many devices, and the rollout of third-generation mobile phone networks. Also, the prominence of fixed wireless will increase during this period and fixed wireless margins will start to exceed those of mobile. More engineers will be required to develop the complex software needed to handle the increasing number of cells and automatically optimize the network.

For the operator, the key changes will be the provision of postmaster and other intelligent functions, the addition of third-generation networks, enhancements to second-generation networks, and the start of a transition to IP networks.

For the end user, this period will see a gradual increase in data capabilities in mobile devices, the use of mobile within the office, and some enhancements in simplicity as BlueTooth and packet data networks enable them to seamlessly retrieve data from preprogrammed points in the office or home network. These will be useful advances but still a long way from meeting their real requirements.

17.3.2 2005–2010

17.3.2.1 Key events

- Homes will start to deploy BlueTooth nodes.

- Communicator devices will be able to work on these in-home networks.

- Postmaster functions will become widespread such that users will be able to have messages forwarded to them in the manner they wish, whatever network to which they happen to be connected.

- E-mail and telephone systems will be linked together using mobile IP standards.

- Video communications will start to become widespread and make up around 2% of calls.

- Broadband connections to homes, perhaps using wireless, will become common to high-end customers.

- Data rate requirements will reach around 10 Mbps.

- Speech recognition will become a standard means of communicating with machines.

- Machine-to-machine communication protocols will be widespread and robust toward the end of this period.

- Core IP networks for cellular systems will become commonplace.

- Public W-LANs in hot spots will become commonplace.

17.3.2.2 Network structure and user capabilities

The likely structure of the network in 2010 is shown in Figure 17.3. A number of key changes have taken place since the network of 2005, even if the network diagram in Figure 17.3 does not appear overly different. Within the office, mobile phones will now be able to communicate with most networks. The office networks will no longer be connected back to the PSTN, but to a core IP network enabling other networks that also

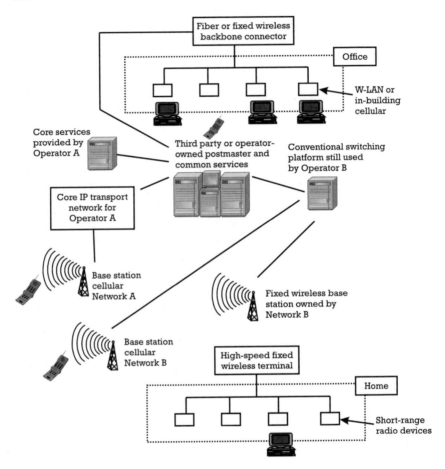

Figure 17.3   Predicted network architecture in 2010.

have a core IP base to communicate in a more seamless manner. This means that the office network can inform the mobile network that there is a message for the user, rather than the mobile device being programmed to search for messages periodically. Not every cellular network has moved over to an IP base, however, and so not all will integrate seamlessly with office networks. This will mean that there will be some segmentation—business users will prefer the more advanced networks and will pay for the service, whereas some residential users will pay less for a lower functionality network.

Radio devices have also entered the home, adding to the in-home networks that some users had already implemented. These will start to work with some mobile devices for the users that pay additional fees for this capability. These users will now have removed their fixed phone. There may, however, still be some difficulties in that the high-speed connection into the Internet may not allow all the functions required for unified messaging and other services.

Third-party intelligent postmaster functions will have started to appear. Coupled to these will be widespread third-party provision of services introduced through the Internet. Not all networks will work with these intelligent postmasters and some operators will still be deploying their own postmasters. Postmasters will be developing rapidly by this point, able to provide very innovative and personalized services to the end user.

Ad hoc networks are now well developed, with PDAs, phones, computers, and other similar devices routinely communicating with each other. Machine-to-machine communications is starting to become more normal, but there are still many legacy machines and still a lack of definition of languages preventing, for example, the refrigerator talking to the PDA in any meaningful manner.

For the manufacturer, the key shift will have been from the manufacture and sale of third-generation networks to software functions like the postmaster, to integrating networks, to IP core network, and also to the provision of additional radio capacity as video communications starts to become more prevalent. Mobile terminals will require increased research and development as functions such as multiple standard support, speech recognition, and ad hoc networking add to the overall complexity.

For the operator, this will be a time of much change where the key differentiators will be the services that can be provided across the network. The operator, however, will be competing with other third parties to provide these services, and there will still be much difficulty over who "owns" the customer. Operators will have split their role into two parts: one part that provides the "pipe," which transfers information from the mobile device, and the other part that provides the services, the postmaster function, and much more.

For the end user, their requirements are now starting to be met. They will be able to have a single phone and a single number and complete unification of messaging if they select an appropriate set of networks and if they pay an additional fee for the provision of this service. They will be able to use video communications, although this will still be somewhat of a novelty. Their machines will start to communicate with them, although they will not be communicating with each other with the full functionality that might be desired.

17.3.3 2010–2015

17.3.3.1 Key events

- Home radio networks will be widespread and most electronic devices will come equipped with BlueTooth capabilities.

- Video calls will increase to as much as 50% of all calls.

- Most homes in developed countries will have high-speed connections, probably based on wireless.

- W-LANs will start to become widely deployed in public buildings, dense urban areas, and malls in such a way that they can be accessed by all users.

- Machines will start talking to other machines.

- All networks except legacy networks will start to use core IP networks with common structure and protocols.

- Fourth-generation cellular networks will start to be introduced.

17.3.3.2 Network structure and user capabilities

The network structure by this time is shown in Figure 17.4, which depicts the same network structure as shown in Figure 17.1; that is, it is the "final" predicted structure for the network for the 20-year period of prediction. By this time we have widespread home networks, ubiquitous high-speed connections to the home in developed countries, W-LANs in most buildings (which can be accessed by most mobile devices), and all key networks using IP core structures, which are able to readily communicate with other networks, enabling full integration at the network level between fixed, cellular, office, and home systems. Bandwidths have risen to the point where video communications is within the cost that most users are prepared to pay, and video communications now comprise as much as 50% of all calls.

Ad hoc networks are now well developed and many machines are now talking to other machines in a meaningful manner. The mobile device is now an integrated phone and PDA, replacing the wallet and becoming the key device carried on the person. Using speech recognition,

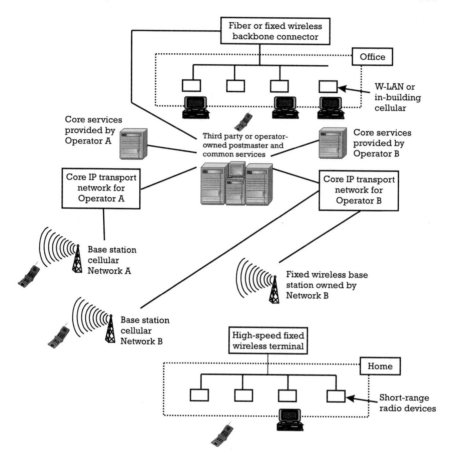

Figure 17.4 Predicted network architecture in 2015.

video screens, and other interfaces, it can be used to control most machines owned by the user and can access most public networks in the world.

Around 2012 we might expect the first fourth-generation cellular networks to be deployed. These will provide even more bandwidth and efficiency and still better data capabilities than third-generation systems. They will also enable better interworking with other networks. They will have enhanced core IP networks based on the lessons learned from third-generation systems.

For the infrastructure manufacturer this will be a period of introducing fourth-generation systems. Additional development will be taking

place on the handset side as handsets become PDAs and "personal electronic wallets." Manufacturers will need to access new capabilities in the
arena of PDA and other areas in order to master this area, and computer
manufacturers and others will also be interested in entering this arena.
The complexity of software will also be increasing rapidly as the devices
become more intelligent and flexible and yet simpler for the user to
operate.

For the operator, now split into two parts, this is generally a continuation of the issues experienced in the previous five years. The
network transport part will be starting to implement fourth-generation
systems and seeking ways of increasing the bandwidth in order to handle
the increased volume of video and data traffic. For the marketing and
service side this is a time of increased standardization with a wide range
of different segments such as banking, travel, and entertainment so that
the "personal wallet," through the network, can access all these different
functions. Operators will also need to be increasingly tightly coupled to
other operators, both of cellular and fixed wireless networks, as well as
owners of in-building networks in order to ensure that all the functionality
required by the user can be provided. Users will be increasingly concerned
with the provision of functionality that simplifies their lives rather than
issues such as QoS, which will be taken for granted. Hence, the operator
that can simplify their lives the most will be the operator of choice.

For the end user this will be a time of much change, as users move
toward the electronic wallet concept. Few will now carry any other
device than their mobile communicator. They will use it for many tasks
that previously were manual, such as the generation of shopping lists.
The capability of the communicator and the other interworking machines
will have significant social implications, in a similar way that the Internet
has changed society. By 2015, most users will be able to communicate
with almost any human or machine using their mobile communicator.

17.3.4 2015–2020

17.3.4.1 Key events

- Data rate requirements will reach around 60 Mbps.

- Most machines in the home will have wireless communications
  capabilities and will communicate with each other.

- Fourth-generation cellular networks will become widespread.

17.3.4.2 Network structure and user capabilities

The network structure at this point is the same as for the previous five years. Indeed, there is little to add from the description of the period of 2010–2015. Data rate capabilities in all the networks will be increasing such that most files can be sent easily to any location. Broadband connections to the home will reach 60 Mbps and will be very widely available around the developed world. In-home networks will provide similar data rates, and in-office networks may reach even higher data rates. Cellular bandwidths will increase as fourth-generation networks become more widely available.

To some extent, this will be a period of consolidation. Most of the foreseeable requirements of the end user have now been met, and the user will have gone through a period of very substantial change. It will take time for these changes to ripple through society and for their implications to become apparent. With users being able to communicate with any person or machine in any way that they want, they will behave in different and difficult to predict ways. These will eventually lead to more requirements—requirements that are almost impossible to foresee today. It will be these requirements that will shape the direction of mobile communications as we move beyond 2020.

By this time, operators are very different from those we have today. A number of them will simply be providers of bandwidth and others will provide services. Manufacturers will have changed significantly, with many new entrants from the arenas of computing and entertainment, and with much consolidation and change.

An interesting point of note is the decreasing number of "key events" predicted as time goes on. In practice, the number of events is likely to be consistent through time, so the fact that the predicted number of events decreases indicates that we are reaching a "prediction horizon" where the techniques used here are no longer able to predict what is likely to be happening 20 years from now. This will come as no surprise. Prediction is always harder, the longer the time horizon—this is simply evidence of that occurring.

In Section 17.4 we use the "day in the life" technique to illustrate the impact of all these trends on the end user. Some of the contributors have already used this approach, and the following section is, in no way, intended to detract from their contribution. Instead, it is intended to sit alongside as another view of the future. What is encouraging, as will be seen, is the similarity of all these visions. What is interesting are the different aspects of future communications on which each of the analyses focuses.

17.4 A vision of communications in 2010

It is May 21, 2010. John Smith is asleep at home. Today he has to take a business trip from New York to San Francisco where he is going to make a presentation to a key customer. He will stay the night in San Francisco and fly back the following morning. The week before, John booked his flight using his personal communicator interlinked to his company's travel agent. The personal communicator has stored these details and has advised John that he will have an early start that day. John has instructed his communicator to wake him at 5:30 A.M. so that he can get to the airport on time.

At 5:20 A.M. the communicator utilizes its BlueTooth capabilities to communicate with the in-home network. The in-home network provides power to a special socket into which the coffee maker has been plugged, slowly turns on the lighting in key parts of the house, and changes the temperature in various rooms—warmer in the bathroom, cooler in other parts of the house.

At 5:25 A.M. the communicator talks to the cellular network to which John subscribes and requests traffic information for the journey to the airport. The communicator sends the start and end address to the network, which returns the optimal route, provided by a third-party Internet traffic service affiliated to the network operator. The communicator stores this information for download to the navigation system in John's car once the communicator is in proximity to the car. The communicator also requests flight information, checking departure times and gate information.

At 5:30 A.M. the communicator plays wake-up music that John has set as his preference until it hears John say, "Alarm off." It then presents John with his itinerary for the day. John gets ready to leave the house, drinks his coffee, and then climbs into his car. The car downloads the journey details from the communicator and provides John with directions for his journey to the airport.

John subscribes to a personal news service provided by his cellular operator in conjunction with Reuters. His communicator downloads the audio signal, comprised of items likely to be of interest to John, and then uses BlueTooth to send this signal to the car audio system, which then plays it back to John. John can request additional information on any story simply by asking his communicator for "more details on previous story." On route, John remembers that he will miss a key television program that evening. As yet, however, there is no simple link into his home network once his communicator has left the house. Instead, he leaves a message for his wife, still asleep, to set the video when she wakes.

Once John has finished listening to his news broadcast, his preferences have been set such that if he is still in the car, he retrieves any messages and listens to them. His communicator talks to the cellular network and asks it to contact the office network and e-mail system to determine whether there are any messages waiting. Once these have been retrieved, John is told of the messages he has and asked whether he wants to listen to them. Using predefined voice commands, John listens to a number of the messages, which have been turned into speech by a function in the network prior to being transmitted to the communicator. He can reply to these e-mails using voice recognition software located in the network that he has already trained, but generally prefers not to do this, especially in the car, as the software sometimes makes recognition mistakes because of the background noise and these can be difficult to spot and correct.

Once John reaches the airport, the communicator provides him with a plan of the route to the gate. At the gate, John has to utilize a smart-card to obtain his boarding pass since his communicator cannot inter-work with the W-LAN within the airport. While he is waiting to get on the plane, John opens his laptop and takes a look at the presentation that he will be giving that afternoon. He notices a missing slide and needs to access the office network to retrieve it. The laptop locates an indoor picocell provided in the airport by the cellular operator and calculates a download time of 45 seconds. John accepts.

The aircraft is one of the newer ones in the American Airlines fleet and has been fitted with a BlueTooth network. John's PDA locates this network and registers its presence. John has already informed the PDA of his preferences when on aircraft, which are to receive e-mail headers in addition to voice-mail messages in text format but not to accept calls because of the high cost still associated with communications on aircraft.

Upon landing, John's PDA locates a high-bandwidth network in the airport and downloads all e-mail and voice-mail and notifies John of their presence. It retrieves the car rental information booked by John, and through the cellular network it informs Hertz of John's arrival. Hertz verifies the electronic signature provided by the communicator and checks with the cellular operator that the communicator has not been reported stolen. It then provides the operator with the bay number in which John's car is located. This is sent to the communicator along with a map to guide John to the bay. Once in the car, as before, driving directions are provided and messages are relayed through the car's audio system.

John arrives at the customer organization and meets his colleagues. The conference room has a projection unit equipped with BlueTooth. John turns on his laptop and uses the BlueTooth link to connect to the projection unit. During the presentation, the customer asks a question that John cannot answer. John quickly uses his PDA to send a voice-mail message to his office and asks for immediate notification of the response. The PDA has not detected any wireless network within the office, but fortunately there is sufficient signal strength on the cellular network. While he is finishing his presentation, his PDA downloads a short presentation that a colleague in the office had assembled to address his question. This is sent to the laptop using BlueTooth and then forwarded to the projection unit.

That evening, in his hotel room, the PDA detects a BlueTooth node and communicates with the hotel network. It downloads the hotel rates for phone calls. When John asks the communicator to "call home," it asks him "audio or video" knowing that for calls home John may use video if it is within the price range he has preprogrammed into the device. John decides on video communications and places the PDA on the desk, sitting himself in front of it. His wife answers the call and they talk about the day. His wife tells him that their son, who is traveling in China, has sent a video postcard, which she downloaded in just a few seconds on their new 20-Mbps wireless Internet connection—would John like her to forward it to his mobile? John knows that hotel room downloads can be quite expensive and so decides to wait until he returns home to see the video clip.

If we examine John's day, we might initially think that he has gained much over the communications capabilities that he had available to him in 2000. Certainly, those trying to retrieve voice-mail with the mobile phone in one hand while trying to drive the car and simultaneously navigate with the other hand will find their life much improved. Almost all of what John achieved, however, is possible with today's technology, with the exception of communication between the mobile and the house and between the mobile and the car. All that has happened is that these capabilities have become more ubiquitous and better integrated.

There is also much that John cannot do that he might have liked to do. He was unable to communicate with his house once outside of that environment. He was unable to talk to W-LANs outside of his own office. He found video communications too expensive in some environments. He could not seamlessly access messages left at his home. He was not directly informed of incoming messages; his communicator had to seek these out for him. He was unable to communicate directly with any site on the Internet; rather, services such as the seamless provision of directions were provided by the cellular operator in conjunction with their chosen partner.

Despite all these problems, this vision represents relatively rapid acceptance of technology among the population. It represents a significant change in behavior, one that places faith in an electronic device to run much of the important events in an individual's life. Many would perceive these changes to be very substantial in a 10-year period.

17.5 A vision of communications in 2020

It is May 21, 2020. John Smith is asleep at home. Today, just as he did 10 years earlier, he has to take a business trip from New York to San Francisco where he is going to make a presentation to a key customer. These trips are now much less frequent because John tends to make more use of videoconferencing rather than attending in person, but this time the customer is sufficiently important that he wants to show this by doing things the "old-fashioned way." He will stay the night in San Francisco and fly back the following morning. He notes that much is the same as 10 years before, but some things have improved. Just as before, John booked his flight using his personal communicator interlinked to his company's travel agent. The personal communicator has stored these details and advised John that he will have an early start that day. It has checked expected travel conditions and calculated that John will need to get up at 5:30 A.M.; however, early the following morning it will recheck traffic conditions, recheck the flight schedule, and determine exactly when John needs to wake up.

At 5:20 A.M. the communicator utilizes its BlueTooth capabilities to communicate with the in-home network. It prewarms the water to the showerhead and starts to prepare a breakfast for John. It confirms with

John's preferred travel site the route that John will take and downloads this to the car, checking that the car has sufficient fuel and that its diagnostic systems have not detected any problems.

At 5:30 A.M. the communicator plays wake-up music that John has set as his preference until it hears John say, "Alarm off." It then presents John with his itinerary for the day. John gets ready to leave the house, drinks his coffee, and then climbs into his car. The communicator starts the car just before the journey so that the interior is warm, opens the garage door as John climbs into the car, and closes the garage door as John's car turns out of the drive. It locks all the doors in the house and confirms with the house control system that the house is secure.

John still subscribes to a personal news service, although he now subscribes directly with a number of news agencies rather than through the cellular operator. His communicator visits these sites and retrieves his daily summaries. For playing back in the car, these are still in audio format, but if John finds any of them particularly interesting, he may say, "Retrieve video," and the video image will be downloaded and stored on the communicator for subsequent viewing. His preferences are for all sports items to be handled in this manner—John will view them when he is on the plane. En route, John remembers that he wanted to get the grass cut today. He calls back to the house control unit and instructs it to pass a message to the robot lawnmower to cut the grass today in addition to its weekly schedule. The robot warns him that although it has sufficient fuel for this cut, it will need refueling before it can complete any subsequent cuts.

While John is conducting these activities his communicator notifies him of an important e-mail message. This is in the form of a video clip. The communicator knows that John is unable to view video clips in the car and so plays the audio only. John listens to the audio and dictates an immediate response. Speech recognition has improved much in the last 10 years, and John is confident that not only will it recognize his speech correctly, but it will also enhance the syntax and grammar prior to transmission.

Once John reaches the airport, the communicator provides him with a plan of the route to the gate. The communicator uses the W-LAN system to link into the airline's network and electronically check in, presenting John with his seat number. The communicator will authenticate John as he passes through the boarding gate using BlueTooth. John still gets in line early—carry-on baggage space on the aircraft has not improved significantly over the last 10 years.

John remembers that 10 years ago he brought his laptop as well as his communicator. Now he sees little point in the laptop. There is no need for a keyboard since speech recognition provides the primary interface, although there are some occasions where John desires a little more privacy. The screen on his communicator is sufficient for viewing most images, and besides, hotel rooms provide large screens that his communicator can utilize, communicating with them using BlueTooth. He can access any information he desires from almost any location with the high-speed download capabilities of his communicator, and the device can store many of his key files on miniaturized storage units. While in the lounge he takes a quick look at his presentation on the communicator screen and makes some small changes.

Although baggage storage has not improved much, flying has become much more pleasant. The seat back has a large display, which John can use for many purposes. He can use this as a screen for his PDA, which he does initially to watch his prestored news video clips. He can play games based on his PDA, such as chess, interactively if he so desires. Communication from plane to ground has become much less expensive so John is happy to receive incoming calls and e-mail, although he still chooses not to make video calls while airborne. John can select from a range of films provided by the airways, using his PDA to present him with choices most likely to be of interest.

Arriving at the airport, John's communicator informs him that if he leaves now for the customer meeting he will arrive 15 minutes early. Would John prefer to wait at the airport? John asks the organizer for a list of coffee bars at the airport and the prices of a latte. The organizer finds one that meets John's preferences nearest to him, provides him with directions, preorders his coffee and debits his electronic purse built into the organizer. His coffee is waiting for him when he arrives. While drinking the coffee, John requests a refresher on the history of the company he is visiting and watches the video clip downloaded over the airport W-LAN system. His communicator informs him when it is time to leave and directs him straight to where his hire car is waiting for him.

When John arrives to meet his customer, he talks with him about the changes that have taken place in communications since his last visit. They compare the features and functionality of their latest organizers and some of the latest advances. John's communicator integrates with heart rate and blood pressure monitors interwoven in the fabric of his shirt and monitors the information for irregularities. His colleague has a miniaturized version built into his wristwatch. This makes use of retinal projection to produce what appears to be a full-sized image when raised up to eye level.

In the hotel room, the communicator downloads the room service menu, filters it, and presents to John a list of options in the order that it thinks he will prefer. John, however, decides to go out for a meal and asks his communicator to assemble a list of nearby restaurants that have availability in his normal price range. He looks at video clips of the interior of the restaurants and selects his preferred one. A booking is immediately made and the directions downloaded to his communicator—they will be passed to the car when he gets close. Before going out, John makes a number of calls, all video calls, to his home and colleagues. He then looks at his video-mail messages and replies to some.

Upon his return to the hotel room, his communicator dims the lighting, locks the door, and prepares the room for John to go to bed. It informs John of his schedule the following morning and suggests an appropriate wake-up time. John accepts. Before going to sleep he reads some more of a recent novel displayed on the screen of his communicator.

John has now achieved his vision of being able to talk to any person or machine in any manner that he likes. His personal communicator is now controlling much of his life, including the environment around him as he travels, and filtering the information that he is presented with. The communicator has become truly indispensable in helping John organize his life.

Although much is now enhanced in the way that John can communicate between 2010 and 2020, this is perhaps less of a paradigm shift than between 2000 and 2010—the period when the communicator started to organize John's life. All that has really happened since then is that the communicator has gotten better at talking to other devices in many diverse environments.

17.6 Key uncertainties

Throughout this book we have noted a number of areas of uncertainty. These are generally areas where we have noted the need for particular developments or evolutions but have not been certain about the exact form that these developments would take, or have not been sure if the economics or politics would allow these developments to take place.

Although we discussed the issues and problems associated with these developments in the earlier part of this chapter, here we summarize these issues and look in more detail at the implications of the uncertainty.

These key uncertainties identified in this book are the following:

▸ The traffic levels and services that will actually be required by users and for which they will be prepared to pay;

▸ The extent to which unlicensed spectrum might become congested, rendering the use of devices in this band problematic;

▸ Possible problems created by health scares and environmental issues;

▸ Whether W-LAN, PAN, or in-building cellular networks will be widely realized, especially given the low penetration to date, and the actual form that these networks will take;

▸ Whether mesh technology will enable high-speed fixed wireless connections to the home;

▸ Sizes of files and maximum data rates that will be required;

▸ How quickly the machine-to-machine language is developed and how flexible this proves to be;

▸ Whether there will be sufficient collaboration to realize a common core network protocol allowing disparate networks to communicate with each other.

The contributors identified the following additional uncertainties:

▸ Whether it was possible to predict the applications and services that would be used 10 and 20 years hence. On balance, most thought that an estimate could be made of some of the services that would be popular.

▸ Whether the communicator MMI would improve sufficiently to enable mobile data applications. Most thought it would.

▸ Whether the cellular operator would be truly interested in providing data services as the revenue from these fell. Most felt they would, although few looked into this in detail.

› Whether mobile wireless access to the Internet would surpass fixed access. On the one hand the convenience of mobile access is greater, on the other the data rates and MMI of fixed access bring many advantages. Most felt that mobile access would be highly significant.

› Whether wide area cellular networks can deliver broadband data services. Most thought not, depending on the definition of broadband. Most saw video communications as a possibility so implicitly expected data rates of, say, 100 Kbps as a minimum.

We now look at each of the former issues in more detail (the uncertainties identified by the contributors have already been dealt with at the start of this chapter).

17.6.1   The traffic levels and services that will actually be required by users and that they will be prepared to pay for

We have predicted that users will want a wide range of services, including video calls, viewing of broadcast news, and transfer of large files. In estimating the frequency of invocation and length of use of these services we have had little more than educated guesswork to rely on. We will more than likely be incorrect. If the demand for these services is much lower than we have predicted, then there will be fewer drivers for network operators to enhance their networks and we may not see the development of some of the services, such as video calls, in the time scales that we have predicted. In the extreme case, it is possible, but unlikely, that throughout the period under consideration here wireless communications remains a mostly voice dominated activity with some simple text file transfer. The alternative is that we have underestimated the demand for services—something that was very much the case in the last 20 years. The implications of this are that network capacity will have to be greater than predicted. We have already noted that this is problematic in cellular networks away from urban areas, and this fact will probably limit the services that can be supplied regardless of the requirement. In urban areas and buildings this will probably hasten the development of microcells and picocells, regardless of problems of backhaul. We have already noted that generally few technical breakthroughs are required to realize much higher bandwidth in urban areas; therefore, if the demand increases, the timing will change, bringing forward the advent of high-capacity video-calling capability and much more. So, probably, the key issue here is one of timing.

17.6.2 The extent to which unlicensed spectrum might become congested, rendering the use of devices in this band problematic

We discussed in Part III that many wireless communications devices plan to use unlicensed spectrum and that this might lead to unanticipated interference between them. In developing the future vision, however, we assumed that this problem would be overcome in some manner. Here we consider what will happen if it becomes a serious issue. Essentially, excessive interference will limit, or prevent, the operation of wireless communications in these bands. The key devices that we have predicted will use the band are W-LANs and PANs. If interference does become excessive, then these devices will need to be moved to different, dedicated frequency bands. Technically this is unproblematic. Practically, however, it will require the recall of all existing devices and their replacement with new devices. Because cellular phones may also interwork with W-LANs and PANs, these will also need to be recalled if they are to interoperate. The cost of doing this will be substantial, and many will not bother to change their cellular phone and rather accept the restriction in usage for some years. Eventually, new bands will be found and devices updated, but this could easily take five years. Hence, this issue might delay the future vision, perhaps by five years.

17.6.3 Possible problems created by health scares and environmental issues

We predicted earlier that health concerns would lessen as time progressed. However, the opposite might happen. This would result in some of the population either refusing to use wireless devices, or limiting their use. This would certainly reduce penetration levels, but the remaining users might drive requirements as predicted in this book, so the future might look the same, but with fewer users. Innovative ideas will be utilized to reduce concerns; for example, transmitters with protective backing could be located around the users' waists so that radiation directed into the user would be much reduced. Protective innovations will satisfy most users. So, in summary, this seems likely only to change penetration levels, not timing.

17.6.4 Whether W-LAN, PAN, or in-building cellular networks will be widely realized, especially given the low penetration to date, and the actual form that these networks will take

Our vision of the future has called for the extensive deployment of W-LANs and PANs, particularly for communications with machines.

As discussed earlier, however, the penetration of W-LANs to date has been small. It is reasonable to question whether there will indeed be widespread deployment. In Chapter 16 we outlined our prediction that widespread deployment would occur, but here we look at the implications in the case that extensive deployment does not occur. We can differentiate between W-LANs and PANs since these are quite separate initiatives that will or will not succeed independent of one another.

In the case of W-LANs we have noted the need for these to provide communications within the office environment, allowing the same communicator to be used within the office as well as outside, and allowing the download of large amounts of information when in the environment of the W-LAN. If W-LANs are not installed, the key problem will be the integration of the wireless communicator with the office phone system. In the worst case, these will stay separate, like they are today, and users will have to retain a mobile number and an office number. We might expect, however, that cellular operators will seek to exploit this void with in-building cellular networks, or that initiatives such as dual-mode cellular-cordless phones might reemerge to meet a basic requirement. This will enable the functionality predicted here, with the exception, perhaps, of high data rate capabilities. The latter will reduce the frequency of video transmission and the size of files that can be handled.

We have predicted that PANs will have a key role in providing communications within the home and with a range of machines. We have also noted that there is currently no alternative to BlueTooth. The failure of this initiative would be a severe blow to our vision of the future. Almost all machine communications would not occur, applications like utilizing the car navigation and stereo system would become much more expensive and thus less widespread, large file download would be difficult, short-range applications such as gate check-in at the airport would be difficult, and much more. This is a key sensitivity and one that needs to be carefully monitored. Lack of appearance of PANs would render the future much less attractive than described in the earlier parts of this chapter. A delay in the appearance of PANs would equally delay the arrival of the future we have envisioned.

17.6.5 Whether mesh technology will enable high-speed fixed wireless connections to the home

We noted earlier that users might require data rates in excess of 10 Mbps to the home and that this could be provided with fixed wireless mesh technology. Mesh technology, however, is not proven and may be delayed, it may be more expensive than anticipated, or it simply might not prove viable. The latter is the most problematic. In the case that it is not viable, other fixed wireless technologies, such as TDD-TDMA solutions (Section 3.2), might be utilized, but these would be more expensive. More expensive means lower penetration, which means that the vision of the future will be restricted to a smaller subset of users. Delayed means delaying a vision of the future when users can have video communications and other high data rate communications from the home. In practice, the impact of these uncertainties is probably not great. There are other alternatives to providing high data rate communications to the home, including ADSL and cable systems. There are already lower data rate fixed wireless systems that can offer video communications at reasonable quality levels. Hence, although the exact speed and capabilities may vary, this seems unlikely to be an area that will have significant impact on the future.

17.6.6 How quickly the machine-to-machine language is developed and how flexible this proves to be

We noted earlier that the protocol for machine-to-machine communications was indeed a difficult problem and that current efforts appeared to be somewhat restrictive. If this protocol proves difficult to develop, it will limit what can be achieved with machine-to-machine communications; for example, the mobile communicator might not be able to download directions to the car simply because the car cannot understand what the mobile is trying to tell it. The issues here are basically a subset of those discussed above when looking at whether PANs will appear. The lack of a versatile language will significantly impact the development of the future, especially in terms of the capability of the mobile communicator to become a truly indispensable device for the end user. We might expect that this problem would become apparent and that a language would ultimately be developed, but with some delay, which would in turn delay the vision of the future as outlined in this chapter.

17.6.7 Whether there will be sufficient collaboration to realize a common core network protocol allowing disparate networks to communicate with each other

We have predicted that all networks will have IP cores able to communicate with each other to route calls and exchange information about the user. It may be that a range of different network cores exists or that many networks utilize different protocols from IP or that a common higher layer protocol for the routing of calls is not developed. This would also be a major issue for realizing the vision laid out here. Without this interoperability, fixed networks, W-LANs, and, to a lesser extent, PANs will not be able to interoperate seamlessly. This will result in message routing being less than optimal, so that the user may not be able to receive calls when they would like to or may not be able to filter calls in the manner they would desire. The net result of this will be less utility and the inclination of users to have all their services delivered by a single operator able to perform this integration. Eventually, protocol converters and public pressure will overcome this problem. Hence, this may result in a slightly less rich future than envisioned, but this will not be the most serious of the uncertainties.

17.7 The future in two pages

This book contains a substantial amount of information spread across 17 chapters. Distilling this information into a two-page summary is challenging, but does allow us to highlight the key issues we have identified.

We expect the future of wireless communications to be a very bright one. Almost everyone in the developed world will carry a wireless communicator that they will use many times during the day. They will perceive immense value from the services that this communicator provides and this will be reflected in increased ARPUs, which in turn will drive the industry with increased growth and investment. Enormous expenditure will occur as networks are constructed outdoors, in offices, in public spaces, and in the home. Wireless capabilities will be embedded into almost all devices—even clothing.

Looking at networks, we do not expect to see one single radio solution providing coverage everywhere. Instead, there will be a multiplicity of different networks: cellular in the wide area, W-LAN in hot spots, public buildings, and enterprises, simplified W-LANs in the home, and dynamically formed PANs as people move and interact. Fixed wireless will be one solution to deliver high-speed access to the home and the office environment. Some networks will predominantly carry traffic between machines, others will be more focused toward humans.

A multiplicity of networks already exists today. What will be different in the future is that all these networks will work together in a manner not currently possible. This will be achieved in two ways. First, devices will be multimodal and so will be able to work on almost any network that they discover. This will become economically viable as advances in baseband electronics reduce the cost premium of complexity. Second, the networks themselves will be able to interwork through the sharing of a common IP-based core and the development of intelligent call routing functions able to deliver calls to subscribers wherever they are in an appropriate manner.

These changes will provide the underlying framework for an extraordinary change in social behavior—an information revolution. This will be driven by the availability of communicator devices able to perform a wide range of functions with which people can interact in a simple and intuitive manner. This will be enabled through the rapid development of applications and content, which will occur as wireless networks move to an Internet model whereby applications can be created at the edge of the network and downloaded to any entity connected to the network. Improved interfaces on the communicators, including large color screens, speech recognition, and more, will increase their capability to handle complex content. New means of communication, including video-phones and high data rate transmission, will drive social changes such as working at home. These, in turn, will lead to new applications and devices. The evolution of wireless really will change the world as we know it, as reflected in all the "day in the life" scenarios presented in this book.

Of course there are risks and uncertainties involved in painting this picture of the future. Operators are likely to encounter turbulent times as their costs increase with auction fees and while the competition from W-LANs may cause them problems. Virtual operators, global brands, and new entrants will change the face of the industry as we know it today. The environment will be confused for many years as new entities enter and many new concepts are tried and fail. The end result, however, is so obviously advantageous and the willingness to pay for this so clear that failures will be tolerated in the quest to be a leader in the new connected world.

In summary, we are on the verge of a dramatic change in the way that wireless communications impacts on an individual's world. This will not come about through technological breakthroughs but through network integration and network models that enable rapid development of diverse applications by multiple entities and through devices that can enable simple interaction with these applications.

17.8   What should we be doing now?

We have now reached the end of our journey. We have covered a lot of ground, understanding what the user requires, what current technology can do, what the key constraints and drivers for the future are, what a range of industry experts expect, and finally a derivation of what might happen based on an analysis of the key issues and areas of uncertainty. This was encapsulated in the visions of 2010 and 2020. This is as much as we can do today in terms of predicting the next 20 years. Now it is time to turn from prediction to action in order to make this future happen. This section considers some of the things that we should be doing now in order to enable this future vision.

17.8.1   Research programs required

In general, we have noted that there are few technical breakthroughs required to realize our vision of the future. Most of John's day in 2010 or even in 2020 can be achieved with current technology—what is lacking is the integration of multiple disparate systems into one simple entity. There were, however, a number of areas identified where research appears to be required, including:

- *Location technology.* Although there are many options for locating a user, all have their drawbacks. A single simple solution to this area is needed.

- *Man-machine interface.* We have noted that a poor MMI may be a key reason for not using wireless devices. Research into areas such as retinal projection, speech recognition, and fold-out displays is required to ensure devices remain usable.

- *Battery technology.* As devices become more complex, they will consume more battery power. Batteries are already the limiting factor in many areas.

- Mesh networks are needed for high-speed connections to the home.

- Techniques are needed for better sharing of unlicensed spectrum given that many of the networks discussed here will seek to operate in unlicensed bands. New ways to avoid interference between uncoordinated devices would enable more networks to be deployed.

• Means are needed for converting content depending on the device and environment that it will be delivered to in an efficient manner.

• Research into health and social issues are needed to quell concerns that the use of mobile devices will lead to undesirable outcomes.

• Better approaches to mobile IP addressing are needed that do not require redirection and encapsulation. With the expected consolidation of e-mail and voice-based systems, efficient means of addressing and routing for both will be required.

Of course, there are many other areas of research that will continue, such as more powerful DSPs, adaptive antennas, and much more. Their omission from the list above is not to diminish their importance. Research in these areas is already under way and we expect it to continue, and hence, they do not need special attention.

17.8.2 Standards issues

Because we have noted that the key to the future is integrating different networks together, we might expect that there is a greater dependency on standards than there is on technology research. Many standards activities are under way. Additional required standards might include the following:

• Increased strength in the fixed wireless standardization area, perhaps focused on areas such as mesh technology;

• A standard for the exchange of information between communicator devices and cars. Although BlueTooth will probably be used as the physical layer, higher level protocols would also be needed;

• Integration of different networks requires substantial standards work to enable, for example, an IP-based office LAN to communicate with a cellular network. A common language would be ideal; a way of translating between different languages is more likely. This may require both a new standard and modifications to many existing standards. This cannot be stressed enough: this "integration" work will require enormous standardization effort;

• Machine-to-machine protocols, although partly covered within the BlueTooth profiles, may need much additional work;

> Security systems able to work across multiple devices and environments.

17.8.3 Spectrum issues

More spectrum is always a welcome way to increase capacity, and hence, any additional allocations would aid the industry. Specific areas where further allocations or harmonization may be required include the following:

> Increased 3G spectrum allocations to enable cellular systems to provide greater data rates. Also, harmonization of 3G bands across the world to minimize the need for multiple RF front ends;

> Spectrum for in-home systems, perhaps at 5 GHz, on a worldwide basis;

> Increased unlicensed spectrum, including enlarging the band at 2.5 GHz to minimize the possibility of excessive interference.

17.8.4 Regulatory issues

In general, most of the discussions within this book have commented on the need for "light-touch" regulation as opposed to specific intervention. It is clear that excessive government intervention might harm growth. Of particular interest were the security issues around enabling access to personal data for those network operators that needed it to deliver appropriate services while at the same time keeping this data secure. Ensuring socially valuable content and universal access was also briefly mentioned by the contributors.

17.8.5 Individual skills

Key to this future environment is the integration of different networks. So, while there is a continued need for designers and research engineers, there will be a developing need for individuals who can enable different types of networks to work together. These "integration specialists" will need to be able to understand different types of networks and be well versed in protocol theory in order to be able to develop the unifying language. They will be able to both write the standards and integrate the networks. Other areas where skills will be valued more in the future than now include the following:

- Applications developers;

- User interface specialists;

- Security specialists.

17.8.6 Shared vision

The future is the realization of the separate visions of those in a position where they can have influence. The future happens more quickly, and more coherently, when that vision is widely shared. Many readers will be involved in the wireless communications industry. If you believe in this vision of the future, go ahead and make it happen. If you disagree in some areas, as many will, understand and promulgate your alternative. In 2010 and 2020 we can look back and see just how well we did both in prediction and concerted action. I think we might be pleasantly surprised at how prophetic we were.

List of acronyms

3GPP   3G Partnership Program

AC   authentication center

ADSL   asymmetric digital subscriber line

AMPS   advanced mobile phone service

ARIB   Association of Radio Industries and Businesses

ARPU   average revenue per user

ASCI   advanced speech call items

ATIS   Alliance for Telecommunications Industry Solutions

ATM   asynchronous transfer mode

BPSK   binary phase shift keying

BRAN   broadband radio access networks

BTS   base transceiver station

CAMEL   Customized Applications for Mobile Enhanced Logic

CD   compact disc

CDMA   code division multiple access

CDR   call data record

CEPT   Conférence Européenne des Administrations des Postes et des Télécommunications

CIR   carrier-to-interference ratio

CLASS   customized local access supplementary services

CN   core network

COO   cell of origin

CSMA/CA   carrier sense multiple access with collision avoidance

DAMPS   digital advanced mobile phone system

DECT   digital enhanced cordless telephone

DSL   digital subscriber line

DSP   digital signal processor

DSSS   direct sequence spread spectrum

DVD   digital versatile disk

EDGE   enhanced data rates for global evolution

ETSI   European Telecommunications Standards Institute

FCC   Federal Communications Commission

FDD   frequency division duplex

FDMA   frequency division multiple access

FEC   forward error correction

FH   frequency hopped

FM   frequency modulation

FTTC   fiber to the curb

FTTH   fiber to the home

GPRS   general packet radio service

GPS   global positioning system

GSM   global system for mobile communications

HFC   hybrid fiber coax

HDSL   high-speed digital subscriber line

HDTV high-definition television

HLR home location register

HSCSD high-speed circuit-switched data

HSPA high-speed packet access

HTML hyper-text markup language

IEEE Institute of Electrical and Electronic Engineers

IMT 2000 International Mobile Telecommunications for the Year 2000

IMTS improved mobile telephone service

IP Internet protocol

IPR intellectual property rights

IRR internal rate of return

ISDN integrated services digital network

ISI intersymbol interference

ISM industrial, scientific, and medical

ISP Internet service provider

ITU International Telecommunication Union

LAN local area network

LCD liquid crystal display

LMDS local multipoint distribution system

LOS line of sight

MAC medium access control

MAP mobile application protocol

MIMO multiple input multiple output

MMDS microwave multipoint distribution system

MMI man-machine interface

MoU   minutes of use

MoU   memorandum of understanding (when used in the context of operators)

MP3   Moving Picture Experts Group, Audio Layer 3

MPEG   Moving Picture Experts Group

MPEG-4   Moving Picture Experts Group Standard #4

MSC   mobile switching center

MTS   mobile telephone service

MVNO   mobile virtual network operator

NMT   Nordic Mobile Telephone

NPV   net present value

O&M   operations and maintenance

OFDM   orthogonal frequency division multiplexing

OSI   open systems interface

PAN   personal area network

PBS   personal base station

PBX   private branch exchange

PC   personal computer

PCM   pulse code modulation

PDA   personal digital assistant

PDN   public data network

PIM   personal information manager

PMP   point to multipoint

PMR   private mobile radio

POP   population

POTS   plain old telephone service

PSTN   public switched telephone network

PTO   post and telecommunications operator

QAM   quadrature amplitude modulation

QoS   quality of service

QPSK   quadrature phase shift keying

RA   Radiocommunications Agency (in the United Kingdom)

RCC   radio common carrier

RF   radio frequency

RTS   request to send

SDH   synchronous digital hierarchy

SIM   subscriber identity module

SIR   signal-to-interference ratio

SME   small and medium enterprise

SMR   specialized mobile radio

SMS   short message service

SNR   signal-to-noise ratio

SoHo   small office/home office

SS7   signaling system 7

TACS   total access communications system

TDD   time division duplex

TDMA   time division multiple access

TETRA   terrestrial trunked radio

TIA   Telecommunications Industry Association

UHF   ultra-high frequency

UMTS   universal mobile telecommunications system

UWB   ultrawide band

VCR   video cassette recorder

VDSL   very high-speed digital subscriber line

VLR   visitor location register

VoD   video on demand

VOFDM   vector orthogonal frequency division multiplexing

VoIP   voice over Internet protocol

VPN   virtual private network

WAP   wireless application protocol

WASP   wireless access service provider

WCDMA   wideband code division multiple access

W-LAN   wireless local area network

WLL   wireless local loop

About the authors

William Webb graduated in electronic engineering from Southampton University with a first class honors degree and all top year prizes in 1989. In 1992 he earned his Ph.D. in mobile radio, and in 1997 he was awarded an M.B.A., both from Southampton University, United Kingdom.

From 1989 to 1993 Dr. Webb worked for Multiple Access Communications Ltd. as the technical director in the field of hardware design, modulation techniques, computer simulation, and propagation modeling. In 1993 he moved to Smith System Engineering Ltd. where he was involved in a wide range of tasks associated with mobile radio and spectrum management. These included devising the U.K. spectrum pricing proposals and writing the modifications to the GSM standard to introduce group and broadcast calls. In 1998 he moved to Motorola where he is a director responsible for strategic direction and vision across the entire communications portfolio.

Dr. Webb has published more than 50 papers, holds four patents, was awarded the IERE Premium in 1994, is a member of the IEE where he sits on the Editorial Advisory Panel, and is a senior member of the IEEE. He is the coauthor, with L. Hanzo, of *Modern Quadrature Amplitude Modulation* (Wiley and Sons, 2000) and the author of *Introduction to Wireless Local Loop* (Artech House, 2000), *The Complete Wireless Communications Professional* (Artech House, 1999), and *Understanding Cellular Radio* (Artech House, 1998). He is currently based in Chicago, Illinois.

Mark Birchler has been with Motorola for 17 years. He received his B.S. in 1983 from the University of Minnesota and his M.S. in 1988 from the Illinois Institute of Technology. Mr. Birchler has been involved in research for his entire career, with key projects including digital microwave technology, GPS prototype development and field testing, iDEN physical layer modulation and algorithms, iDEN/GSM location technology, and fixed wireless and EDGE systems. Currently, Mr. Birchler manages wireless access and physical interface research in Motorola Labs. He has served as Motorola's primary technical representative to the FCC on location technology issues. He has 17 issued patents.

Larry Marturano leads the user-centered research group within Motorola Labs' Application Research Laboratory located in Schaumburg, Illinois. This multidisciplinary team identifies next-generation applications and services through the study of social, environmental, and technological trends. He holds a Ph.D. from Northwestern University and M.S. and B.S. degrees from the University of Illinois at Urbana-Champaign, all in electrical engineering. In his 14-year Motorola Labs career, Dr. Marturano has researched and published in such diverse topic areas as divided attention and mental workload, speech quality and perception, error control coding, wireless system coverage, and integrated circuit design. Dr. Marturano is a member of the Association of Computing Machinery and the IEEE. He holds four U.S. patents.

Michel Mouly graduated from the Ecole Polytechnique and the Ecole Nationale Supérieure des Télécommunications in Paris, and spent his career to date in the field of digital cellular telecommunications. He was actively involved with the standardization of GSM from 1985 to 1998, particularly in the specification of the signaling, security, and radio transmission standards. Mr. Mouly chaired the sub-group of the GSM standardization committee dealing with radio interface signaling (SMG3/WPA), and then chaired the architecture and signaling committees (SMG3, SMG3/WPA,

and SMG3 system architecture groups) within ETSI/SMG. He has worked for France Telecom and NORTEL Networks. He is now a private telecommunications consultant, working in particular in the areas of GSM and UMTS. Mr. Mouly has authored, with M.B. Pautet, a book on GSM, *The GSM System for Mobile Communications*, which can be ordered on-line from http://perso.wanadoo.fr/cell.sys.

Tero Ojanperä, Nokia Networks vice-president, is responsible for research, standardization, and technology within IP mobility networks at Nokia Networks, which covers a full portfolio of radio access networks, broadband wireless systems, and core networks. Dr. Ojanperä received his M.Sc. from the University of Oulu, Finland, in 1991 and his Ph.D. from Delft University of Technology, The Netherlands, in 1999. While at Nokia Mobile  Phones in Oulu, Finland, he was a project manager for the wideband CDMA concept development, which later formed the basis for the FRAMES wideband CDMA. Dr. Ojanperä has been a research manager in the Nokia Research Center, Helsinki, Finland, heading Nokia's third-generation air interface research program; a Nokia representative for the UMTS radio interface issues in the ETSI SMG5 and SMG2 standardization committees; a principal engineer at the Nokia Research Center in Irving, Texas, involved in U.S. third-generation standards activities for the cdma2000; and involved in technical/strategic work for Nokia's proposal for the UWC-136. In 1998, Dr. Ojanperä joined Nokia Telecommunications (now Nokia Networks), Finland, as the head of research for radio access systems. In 1999 he was promoted to vice-president, radio access systems research, and later became general manager of Nokia Networks, Korea. Dr. Ojanperä is coauthor of *Wideband CDMA Third Generation Mobile Communications* (Artech House, 1998) and its second edition *WCDMA: Towards IP Mobility and Mobile Internet* (Artech House, 2001). He has authored several conference and magazine papers, and has contributed to several books. Dr. Ojanperä is a member of the IEEE and is vice-chair for the International Communications Conference (ICC) 2001.

Malcolm Oliphant is the strategic marketing manager
for strategy and advanced development in the communications business unit at Tektronix. He has more than
30 years of experience with mobile radio systems, particularly with first- and second-generation cellular and
private mobile radio (PMR) systems. Mr. Oliphant previously was the strategic marketing manager at IFR. He
gained most of his digital cellular experience while he was

with Schlumberger in Munich, where he was involved with GSM signal-
ing test products. He has published over 100 articles in RF and cellular
technical and trade publications, is a visiting professor with the Organization of Electronics Industries (OEI), and has coauthored two popular
books on GSM and PCS. Mr. Oliphant received his B.A., *summa cum laude*,
from Hawaii Pacific University, Kaneohe, Hawaii.

Siegmund Redl, a graduate of the Technische Universi-
taet Muenchen (Munich, Germany), has been working in
the mobile communications industry for more than a decade. He has held positions in product management and
product marketing with Schlumberger Technologies, LSI
LOGIC, and Qualcomm's CDMA Technologies Division.
In the early 1990s, Mr. Redl also participated in ETSI GSM
standardization efforts. He has coauthored textbooks on

GSM and personal communications (*An Introduction to GSM* and *GSM and
Personal Communications Handbook,* both by Artech House) and has written
and presented numerous papers on wireless communications.

Mike Short has spent his 26-year career in electronics
and telecommunications, with the last 14 years in mobile
communications. He was appointed contracts director of
Cellnet in 1989, dealing with multimillion-dollar infrastructure investments and U.K. interconnect agreements.
In 1993 the focus moved to establishing Cellnet's GSM
service. He was elected chairman of the GSM Association
for 1995–1996, and served on its executive board for

three years. The number of customers using GSM-based networks now
exceeds 400 million worldwide. Mr. Short's focus today is on third-
generation (UMTS) cellular and steering BT Wireless's industry rela-
tions/standards from GSM evolution toward UMTS. He also is a member

of the U.K. FEI Board and U.K. SMAG (Ministerial Advisory Group), and was elected Chairman of the U.K. Mobile Data Association for 1999/2000. He also chairs the GSM Association Data Task Force and is a member of the WAP Forum Board.

Frank Yester received his B.S. in 1966 from Pennsylvania State University and his M.S. in 1968 from the University of Pennsylvania. Mr. Yester joined Motorola in 1968 and has been a member of various Motorola research organizations since that time. He has contributed in a number of technical areas including custom integrated circuit design, RF power amplifier modeling, analog and digital microwave radio design, and A/D converters for digitally implemented radios.  In 1988 he became a manager of research where he had responsibility for the development of the RF system design, modulation, RF hardware architecture, and custom integrated circuits for iDEN and other digital systems. Currently, Mr. Yester is the vice-president and director of wireless access and applications in Motorola Labs. He has 18 issued patents.

U.K. Radiocommunications Agency: This work was conducted by 30 experts in the field of wireless communications drawn from around the world. The list of names is confidential as part of the "Chatham House" rules under which the workshop was conducted.

Index

Understanding GPS: Principles and Applications,
Elliott D. Kaplan, editor

Understanding WAP: Wireless Applications, Devices, and Services,
Marcel van der Heijden and Marcus Taylor, editors

Universal Wireless Personal Communications, Ramjee Prasad

WCDMA: Towards IP Mobility and Mobile Internet, Tero Ojanperä
and Ramjee Prasad

Wideband CDMA for Third Generation Mobile Communications,
Tero Ojanperä and Ramjee Prasad, editors

*Wireless Communications in Developing Countries: Cellular and
Satellite Systems,* Rachael E. Schwartz

Wireless Intelligent Networking, Gerry Christensen,
Paul G. Florack, and Robert Duncan

Wireless Technician's Handbook, Andrew Miceli

For further information on these and other Artech House titles,
including previously considered out-of-print books now available
through our In-Print-Forever® (IPF®) program, contact:

Artech House
685 Canton Street
Norwood, MA 02062
Phone: 781-769-9750
Fax: 781-769-6334
e-mail: artech@artechhouse.com

Artech House
46 Gillingham Street
London SW1V 1AH UK
Phone: +44 (0)20 7596-8750
Fax: +44 (0)20 7630-0166
e-mail: artech-uk@artechhouse.com

Find us on the World Wide Web at:
www.artechhouse.com